Migration and Health in Asia

The processes of migration and health are inextricably linked in complex ways, with migration impacting on the mental and physical health of individuals and communities. Health itself can be a motivation for moving or a reason for staying, and migration can have implications on the health of those who move, those who are left behind and the communities that receive migrants.

This volume brings together some of the increasing number of researchers who are studying health and migration in Asia, a continent which is a major exporter and importer of human resources. The essays use both quantitative and qualitative approaches to investigate interdisciplinary issues of health and health-related behaviours in the field of migration. *Migration and Health in Asia* covers major themes including the pandemics of HIV/AIDS and SARS, differential access to health and civil society for migrants and the health of the populations who are left behind.

Santosh Jatrana is a research fellow at the department of Public Health, Wellington School of Medicine and Health Sciences, University of Otago, New Zealand. She is a demographer with particular research interest in the field of social demography, such as socioeconomic inequality and mortality, child health, gender, demography of ethnic minorities, ageing and health, migrant health, and maternal employment and family commitments. **Mika Toyota** is a research fellow at the Asia Research Institute, National University of Singapore. She is a social anthropologist whose research focuses on transnational networks, the geopolitics of borderlands, migration, gender and the changing family in Asia. **Brenda S.A. Yeoh** is Professor of Geography at the National University of Singapore, and the Principal Investigator of the Asian MetaCentre. Her research focuses on the politics of space in colonial and post-colonial cities; and gender, migration and transnational communities.

Routledge research in population and migration
Series editors Paul Boyle and Mike Parnwell

Migration and Health in Asia

Edited by Santosh Jatrana, Mika Toyota
and Brenda S.A. Yeoh

LONDON AND NEW YORK

First published 2005
by Routledge
2 Park Square, Milton Park, Abingdon, Oxfordshire OX14 4RN

Simultaneously published in the USA and Canada
by Routledge
711 Third Avenue, New York, NY 10017

Routledge is an imprint of the Taylor & Francis Group

Transferred to Digital Printing 2008

Typeset in Garamond by Wearset Ltd, Boldon, Tyne and Wear

British Library Cataloguing in Publication Data
A catalogue record for this book is available from the British Library

Library of Congress Cataloging in Publication Data
A catalog record for this book has been requested

ISBN 0-415-36319-5 (hbk)
ISBN 0-415-41656-6 (pbk)

Contents

Illustrations

Figures

Tables

Contributors

Maruja M.B. Asis is Director of Research and Publications of the Scalabrini Migration Center (SMC), based in Manila, The Philippines (http://www.smc.org.ph). She is a sociologist who has long been involved in migration studies in Asia, particularly on topics dealing with migration policies, unauthorized or irregular migration, and the relationship between migration on the one hand, and gender, family relations and social change on the other. Since 1997, she has served as Associate Editor of the *Asian and Pacific Migration Journal*, a refereed quarterly journal, and co-editor of the *Asian Migration News*, a bi-weekly online news service. In line with the manifold programmes of SMC in the Asia-Pacific, her involvement in migration issues also entails engagement with non-government and international organizations in promoting the rights of migrants.

Paul Boyle is Professor of Human Geography at the University of St Andrews, Scotland. He is also Director of the Social Dimensions of Health Institute (a joint initiative between the Universities of Dundee and St Andrews) and Director of the Longitudinal Studies Centre – Scotland which is establishing the Scottish Longitudinal Study. His recent research includes studies of the effect of migration on the relationship between deprivation, health and mortality; health inequalities in Scotland; the effects of family migration on separation and women's economic status; and low fertility in Scotland. He is currently co-editor of the journal *Population, Space and Place* and of the Population and Migration book series published by Routledge. He is a co-editor of *The Geography of Health Inequalities in the Developed World* (Ashgate, 2004), *Migration and Gender in the Developed World* (Routledge, 1999); *Migration into Rural Areas: Theories and Issues* (Wiley, 1998); and co-author of *Exploring Contemporary Migration* (Longman, 1998). He is currently collaborating on a project that will examine the health of 'left-behind' children in four Asian countries.

Angelique Chan is an Assistant Professor of Sociology at the National University of Singapore. She obtained her Ph.D. from the University of

California, Los Angeles and completed a postdoctoral fellowship at the University of Michigan, Population Studies Center. She has been researching ageing issues since 1990 when she began studying living arrangements of older adults in Malaysia and subsequently Singapore. Recently, she has been focusing on health issues among the aged; specifically, gender and socioeconomic differentials in health status among older adults in Singapore and the region. Her published works include academic articles and book chapters that focus on ageing in Singapore and comparative analyses of Singapore with other countries in the region.

Alan Banzon Feranil is a Senior Research Associate of the Office of Population Studies based at The University of San Carlos in Cebu City. A graduate of demography from The Australian National University, Dr Feranil has since been working on health and population concerns in the Philippine setting.

Elspeth Graham is Reader in Geography at the University of St Andrews, Scotland. Her recent research work includes an investigation into social deprivation and recovery after a first heart attack, a study of intergenerational relationships, fertility and the family in Singapore, and an analysis of the geography of low fertility in Scotland. She also has research interests in the philosophy and methodology of the social sciences, and her publications cover philosophical and methodological issues, as well as empirical studies in population geography and health geography. She is a co-editor of *Postmodernism and the Social Sciences* (Macmillan, 1992), and *The Geography of Health Inequalities in the Developed World: Views from Britain and North America* (Ashgate, 2004), and has written numerous academic papers and book chapters on philosophy and methods in human and population geography, population policies, and health inequalities. She is currently working on a project that will examine the health of 'left-behind' children in four Asian countries.

Graeme Hugo is Federation Fellow, Professor of the Department of Geographical and Environmental Studies and Director of the National Centre for Social Applications of Geographical Information Systems at the University of Adelaide. He has authored over two hundred books, articles and chapters, as well as a large number of conference papers and reports. His books include *Australia's Changing Population* (Oxford University Press, 1986); *The Demographic Dimension in Indonesian Development* (with T.H. Hull, V.J. Hull and G.W. Jones, Oxford University Press, 1987); *International Migration Statistics: Guidelines for Improving Data Collection Systems* (with A.S. Oberai, H. Zlotnik and R. Bilsborrow, International Labour Office, 1997); *Worlds in Motion: Understanding International Migration at Century's End* (with D.S. Massey, J. Arango, A Kouaouci, A. Pellegrino and J.E. Taylor, Oxford University Press, 1998); several of the 1986, 1991 and 1996 census-based *Atlas of the Australian People Series* (AGPS,

1989 to 1999); *Australian Immigration: A Survey of the Issues* (with M. Wooden, R. Holton and J. Sloan, AGPS, 1994); *New Forms of Urbanisation: Beyond the Urban–Rural Dichotomy* (with A. Champion, Ashgate, 2004); and *Australian Census Analytic Program: Australia's Most Recent Immigrants* (Australian Bureau of Statistics, 2004). In 2002, he secured an ARC Federation Fellowship over five years for his research project, 'The new paradigm of international migration to and from Australia: dimensions, causes and implications'.

Santosh Jatrana was a postdoctoral fellow at the Asian MetaCentre for Population and Sustainable Development Analysis, Singapore, and is currently a research fellow at the department of Public Health, Wellington School of Medicine and Health Sciences, University of Otago, New Zealand. She is a demographer with particular research interests in the field of social demography, such as socioeconomic inequality and mortality, demography of ethnic minority, child health, health of older persons, migrant health, maternal employment and family commitments, gender and social determinants of HIV/AIDS. She holds a Ph.D. from the Australian National University, Canberra. Her work experience in India, Australia and Singapore covers a wide range of institutional set-ups including academic as well as governmental organizations. She has published numerous academic papers and book chapters on infant and child survival, gender disparity in child health, impact of maternal work on child health, health of female migrant construction workers and health of older adults in Singapore. She is a co-editor of a special issue of *Asian and Pacific Migration Journal* on migration and health in Asia.

Gavin W. Jones is Professor in the Asia Research Institute, National University of Singapore, where he is research team leader on the changing family in Asia. After completing his Ph.D. at The Australian National University in 1966, he joined the Population Council, where he worked first in New York, then in Thailand and Indonesia, before returning to Australia. He was then with the Demography and Sociology Programme at The Australian National University for twenty-eight years, serving as head of programme for six years, and conducting research mainly on South-East Asia. He has conducted joint research with many colleagues in South-East Asia, on subjects as varied as the economic crisis in South-East Asia, marriage and divorce in Malaysia and Indonesia, population and development in Eastern Indonesia, urbanization, and education and human development. He also has a long-standing interest in the formulation and evolution of population policy in the region. He has served as consultant to many international agencies, and has published about twenty books and monographs and some 130 refereed journal articles and book chapters. He is currently Chair of the Council of CICRED (Committee for International Cooperation in National Research in Demography), a Paris-based organization. His general interest in population and development issues has, in recent years, focused on inequality of access to education, the dynamics of population and employment change in the mega-urban regions of South-East Asia, and delayed marriage and non-marriage.

Randall S. Kuhn received a joint Ph.D. in Demography and Sociology in 1999 from the University of Pennsylvania in Philadelphia. His research focuses on the impact of family and community networks, and demographic and social change on individual well-being throughout the life-course. In particular his work on internal and international migration from rural Bangladesh addresses migration's role as a household livelihood strategy affecting old-age support, health, family formation, education and further migration. He spent two years in the 1990s conducting qualitative and quantitative fieldwork as a research fellow at ICDDR,B in Dhaka, Bangladesh, and is grateful for assistance he received from his colleagues at ICDDR,B, the good-natured citizens of Matlab, and the tireless field staff, particularly the late A.T.M. Mahfuzur Rahman. Kuhn currently works at the University of Colorado at Boulder, as a Research Associate in the Institute of Behavioral Science's Population Program, and Adjunct Assistant Professor in the Department of Sociology. His current activities include studies of health and survival in migrant-sending regions of Bangladesh and Indonesia; the effects of HIV/AIDS and migration on older South Africans; and the effects of macroeconomic crisis on health in Russia.

Chris Lyttleton is a Senior Lecturer at the Anthropology Department, Macquarie University, Sydney, Australia. His primary research work is in medical anthropology focusing on the social impact of HIV/AIDS in Thailand and Laos; changing patterns of drug abuse in mainland South-East Asia; and mobility, social change and cross-border disease vulnerability in the Upper Mekong. He has published widely on these topics including: 'Fleeing the fire: transformation and gendered belonging in Thai HIV/AIDS support groups' (*Medical Anthropology*, 2004); 'Relative pleasures: drugs, development and modern dependencies in Asia's Golden Triangle' (*Development and Change*, 2004); 'Sister cities and easy passage, mobility and economies of desire in a Thai/Lao border zone' (*Social Science and Medicine*, 2002); and *Endangered Relations: Negotiating Sex and AIDS in Thailand* (Routledge/Martin Dunitz, and White Lotus Press, 2000).

Parveen Nangia is currently Research Director at the Social Planning Council, Sudbury, Canada and a faculty member at the Laurentian University, Sudbury, where he teaches in the Department of Sociology, and School of Social Work. Prior to this, he was a Reader (Associate Professor) at the International Institute for Population Sciences, Mumbai, India for fifteen years. Earlier, after obtaining his Ph.D. on Child Labour from Jawaharlal Nehru University (JNU), New Delhi, India, he joined JNU as a University Grants Commission sponsored postdoctoral fellow, and later as Assistant Professor. He had also served the Office of the Registrar General, India as Senior Geographer. His diverse research interests include migration and urbanization, population and environment, nutrition and health, reproductive behaviour, and racism. He is author of one book: *Child Labour: Cause–Effect Syndrome* (Janak Publishers, New Delhi, 1987); and a co-author of *Working and Street Children of Delhi* (with Rita Panicker, National Labour Institute, NOIDA, 1992); *Community Participation in RCH Programme: Experiences from Health Awareness Units Project* (with S.K. Singh and T.K. Roy, International Institute for Population Sciences, Mumbai, 2000); and *Atlas of Reproductive and Child Health Status in India* (mimeo, with F. Ram, International Institute for Population Sciences, Mumbai, 2001). Much of his research work is published in the form of chapters, articles and research reports. He was one of the five coordinators of the second National Family Health Survey conducted in India in 1998 to 1999.

Archana K. Roy is currently working as a Lecturer in Geography in Banaras Hindu University, Varanasi, India. Her doctoral thesis was on the 'Impact of male out migration on left behind families and women: a case study of Bihar', undertaken at the International Institute for Population Sciences (IIPS), Mumbai, India (2004). She received a Government of India Fellowship to pursue her doctoral research. Prior to this she undertook postgraduate work in geography and population studies at Banaras Hindu University, Varanasi and IIPS, Mumbai respectively. Her fields of interest include migration and development, the status of women, health and environment. She has published research papers in national journals and contributed to an edited book ('Left behind women in Bihar' in *Migrant Labour and Human Rights in India* edited by K. Gopal Iyer, Krishna Publishers, 2002).

Mika Toyota is a research fellow at the Asia Research Institute, National University of Singapore. She is a social anthropologist whose research interests include transnational networks, gender and migration, tourism development, the geopolitics of borderlands and the changing family in Asia. She has conducted long-term field research (1994 to 1997) on transnational ethnic minorities in borderlands between Thailand, Burma and China. She obtained her Ph.D. in South-East Asian Studies from the University of Hull, UK in 1999 and subsequently lectured at the same university for three years. The author of sixteen academic articles in English and Japanese, she is the first editor of a special issue of the *Asian and Pacific Migration Journal* on migration and health in Asia.

Theresa Wong is a doctoral student in geography at The Ohio State University, Columbus, Ohio. She has published on migration, foreign worker policies, and population and the family, including 'Spaces of silence: single parenthood and the "normal family" in Singapore' (*Population, Space* *and Place*, with B.S.A. Yeoh, E.F. Graham and P. Teo, 2004). She was Research Assistant at the Asian MetaCentre for Population and Sustainable Development Analysis from 2001 to 2003.

Xiang Biao was a postdoctoral fellow at the Asia Research Institute and Asian MetaCentre for Population and Sustainable Development Analysis, The National University of Singapore after completing his Ph.D. (2003 Oxon.) at Oxford, and is currently a researcher at the Centre on Migration, Policy and Society at Oxford University, UK. He has worked on migrants in China, India and Australia, and has published extensively on migration and social change, including *Transcending Boundaries* (Chinese by Sanlian Books, 2000; English by Brill Academic Publisher, 2004) and *Global 'Body Shopping'* (Princeton University Press, 2005).

Keiko Yamanaka, a sociologist, is a Lecturer at the Department of Ethnic Studies, and Research Associate in the Institute for the Study of Social Change, University of California, Berkeley. Since 1993, she has been studying transnational migration and social transformation in Japan, focusing on two contrasting immigrant populations: authorized resident Brazilians of Japanese ancestry, and unauthorized Nepalese. In recent years she has focused on feminized migration in Asia and civil society responses. She has published articles and chapters on these topics in both English and Japanese. She has co-edited 'Gender, migration and governance in Asia' (with N. Piper, *Asian and Pacific Migration Journal*, 2003), and co-authored 'Feminised cross-border migration, entitlements and civil action in East and Southeast Asia' (with N. Piper, forthcoming), a report to the United Nations Research Institute for Social Development.

Brenda S.A. Yeoh is Professor, Department of Geography, The National University of Singapore, and Principal Investigator of the Asian Meta-Centre at the University's Asia Research Institute. Her research foci include the politics of space in colonial and post-colonial cities; and gender, migration and transnational communities. She has published over seventy scholarly journal papers and several books, including *Contesting Space: Power Relations and the Urban Built Environment in Colonial Singapore* (Oxford University Press, 1996; reissued by Singapore University Press, 2003); *Gender and Migration* (with K. Willis, Edward Elgar, 2000); *Gender Politics in the Asia-Pacific Region* (with P. Teo and S. Huang, Routledge, 2002); *The Politics of Landscape in Singapore: Construction of 'Nation'* (with L. Kong, Syracuse University Press, 2003); *Approaching Transnationalisms* (with M.W. Charney and C.K. Tong, Kluwer, 2003); and *State/Nation/Transnation: Perspectives on Transnationalism in the Asia-Pacific* (with K. Willis, Routledge, 2004).

Preface and acknowledgements

This book represents one of the first efforts to bring together in a single volume the work of a growing number of researchers interested in simultaneously studying migration and health in the context of Asia. The studies included illustrate not only the multifaceted links between migration and health, but also the benefits of using a variety of research approaches and methodologies. The findings from these studies are important to the development of policies and interventions that potentially increase not only migrants' health, but also the health of those left behind amidst the increasingly fluid and complex patterns of migration in Asia today.

This volume is based on several papers first presented at an international workshop entitled 'Migration and Health in Asia', held in Bintan, Indonesia in September 2003. The workshop was organized by the Asian MetaCentre for Population and Sustainable Development Analysis at The National University of Singapore, and supported by the Wellcome Trust, UK. Apart from this volume, some of the other papers have also been published as a Special Issue of the *Asian and Pacific Migration Journal* on migration and health in Asia (Toyota, Jatrana and Yeoh (eds), Vol. 13, No. 1, 2004).

From the conception of the workshop to selecting, editing and putting together the papers into a book, our work was facilitated by many individuals who gave generously of their time and assistance. First, we are grateful to the following reviewers for their comments on various chapters of the book: Maruja Asis, Paul Boyle, Gouranga Dasvarma, Elspeth Graham, Socorro Gultiano, Gavin Jones, Yohannes Kinfu, Nicola Piper and Omar Rahaman. We are also indebted to the Senior Administrative Officer, Verene Koh at the Asian MetaCentre for her unstinting help in receiving and organizing the papers both for the workshop and this volume. Theodora Lam, Leong Wai Kit and Theresa Wong provided us with timely research assistance. It is to Theodora Lam that we owe the improved layout of some of the tables, figures and maps. We are also grateful to Mike Parnwell and Paul Boyle, who in their role as Series Editors for the Population and Migration series, provided us with insightful comments on the book proposal, and encouragement in developing the idea for the book. We would also like to acknowledge the prompt help we received from Terry Clague, the Commissioning Editor for Routledge Research, in response to all our queries. To all the contributors of the volume, we would like to express our gratitude for collaborating with us in bringing this book project to fruition.

Santosh Jatrana, Mika Toyota and Brenda S.A. Yeoh
Asian MetaCentre for Population and Sustainable Development Analysis
The National University of Singapore

1 Introduction

Understanding migration and health in Asia

Santosh Jatrana, Elspeth Graham and Paul Boyle

Introduction

The relationships between migration and health are multiple. There is growing evidence, albeit incomplete, that the process of migration and the health of individuals and communities are inextricably intertwined in complex ways, with implications for those who move, those who are left behind, and those who host migrants. One classic conceptualization of the relationship between migration and health is provided by Hull (1979), who points out that the causal link between migration and health can occur in either direction – the health of individuals and communities may be influenced by migration, while the health of individuals and communities may stimulate migration. Thus, at the macro-scale, migration may influence population health, although the effects may be quite difficult to disentangle (Boyle 2004). At the individual level, we know that the relationship between migration and health varies with age (Findley 1988). Younger adults who migrate are generally healthier than non-migrants, while older adults beyond retirement age are more likely to move if they are in poor health and, for the latter group, health is often an important factor which influences the decision to move. Health can also influence temporary moves – a growing literature on health tourism, for example, suggests that the use of better or cheaper health care abroad is resulting in new patterns of migration in many parts of the world, including Asia (Borman 2004).

The effects of the migration experience on health have received a significant amount of attention from researchers who focus on the differences in health status and mortality patterns between migrant groups on the one hand, and migrants and the native-born population on the other. Compared to those they leave behind, international or long-distance intranational migrants are often relatively healthy, and this is commonly referred to as a 'healthy migrant effect'. This comes as no surprise, given that migration is usually selective of those who are in higher socio-economic groups (Boyle *et al.* 1998), but there is evidence that changes in the physical and social environment in which migrants find themselves also has a role to play. On the other hand, migrants may be less healthy than those whom they join, such that migrant health falls between those whom they leave and those whom they join (Marmot 1989), although others show that immigrants often have better health than the host population that they join (Anson 2004; Rosenwaike 1990). This selectivity has implications for the health of communities both at the origin and the destination.

Thus moving itself can have both positive and negative implications for health. On the positive side, migrants (especially those moving from a poor to a rich country) may find better access to health care services in the new home country, and may benefit from better diets and/or a cleaner environment. Since such migrants may carry with them a health deficit from their area of origin, it might be expected that their health would nevertheless be worse than that of the native-born population. However, new migrants may also experience certain advantages arising from their continued patterns of beliefs, practices and social contacts that protect them from the dangers of the new environment, resulting in an 'epidemiological paradox' of better than expected health outcomes (Morales *et al.* 2002).

While migration may offer economic benefits in the form of higher wages or opportunities for social mobility, and health benefits (as described above), some migrant populations are vulnerable to health problems (Hull 1979; Rashid 2002). This vulnerability is the result of a combination of factors, including the psychological stress generated from the process of removal and resettlement, and the problems and difficulties migrants confront in dealing with health problems and the health care culture in the new environment (Bollini 1992; Chung and Kagawa-Singer 1993; Shuval 1993), reduced security in daily life (Sundquist 1994), experiences of alienation and discrimination (Kaplan and Marks 1990), reduced socio-economic status (Harding and Balarajan 2001) and language barriers (Cookson *et al.* 2001). These factors may result in ill-health and psychological impairments among the migrant population (Hull 1979; Fitinger and Schwartz 1981; Jones and Korchin 1982; Kuo and Tsai 1986). Moreover, migration itself compels the migrants to adjust to a new life-style which often brings with it a new set of health risks (Evans 1987). Even immigrants who are securely resettled may be re-exposed to diseases when they return home to visit friends and relatives, or associate with newly arrived members of their ethnic group, and this has implications for the spread of disease (Cookson *et al.* 2001). The plight of irregular or unauthorized migrants, without official permission to enter or remain in a host country, may be compounded by pre-departure conditions such as poverty and armed conflict, with no adequate health care for months or years during the travel itself, or because they have been trafficked or smuggled (Taran 2002). In addition, there is a large and growing literature which considers the mental health impacts of migration for refugees and asylum seekers who often struggle to adapt to the various changes in their environment (Berry 1997).

Given the increase in migration – both voluntary and forced – all over the world and the web of interconnections between migration and health, it might be assumed that robust theories exist which could explain the multiple relationships between migration and health in Asia. This is not the case. In an Asian setting, collaborative research both by social and medical scientists is mainly limited to the field of HIV vulnerability among migrants and/or host societies (e.g. Srithanaviboonchai *et al.* 2002). Much of the research at the intersection between migration and health has focused on the developed world where, for example, the role of health-selective migration in explaining the widening health gap between the rich and poor has been examined (Norman *et al.* 2005). The absence of similar studies in Asia may be due, in part, to the complexities of the relationships involved, but also results from a lack of reliable data linking health outcomes to migration status, and the background and experiences of migrants. However, new data sources in Asia are now providing opportunities for such research, which promises to enhance our understanding of the dynamics of human migration and its long-term health consequences in the region. International and transnational migration seems to be an inevitable feature of globalization. Better understanding of its consequences for the health of individuals and communities, along with a recognition that health considerations can also provide a rationale for migration, is required if we are to formulate policies aimed at mitigating the adverse effects of migration on health.

Why study migration and health in Asia?

In general, there is a paucity of research that recognizes the complex causal nexus between migration and health, and evidence pertaining to migration and health in Asia remains sparse and fragmented. Moreover, there is little empirical research suggesting that explanations of the relationships between health and migration observed in Western settings can be extended to Asian societies. Further research is required in Asia and other non-Western settings in order to gain a better understanding of how the relationships between migration and health are manifest across diverse socio-cultural environments. The importance of examining migration and health in Asia stems from the need to pay closer attention to the varied meanings of health and illness, as well as to the social and cultural contexts of its relationships to migration (Boyle and Graham 2003). Empirical studies in an Asian setting provide a valuable opportunity to extend our understanding of such differences and diversity.

Investigating migration and health as it affects countries in Asia also offers a unique opportunity to develop theoretical understandings because Asian countries are both major exporters and major importers of migrants. Over the past few decades, the mobility of people across borders in Asia has both increased in volume and diversified into multiple streams, with different gender, class or race structures, and degrees of permanence. As the process of migration involves uprooting, displacement and resettlement, it poses a major challenge to ensuring that population movement and resettlement are socially productive. This challenge will become an even greater ethical and pragmatic necessity in Asia given the likely future increase in population movements in the region. The cultural and political diversity and differential levels of economic development among Asian countries mean that migrants crossing borders grapple with a wide range of social, cultural, political and economic issues. The ramifications for the public health care systems of host communities and countries can also be far-reaching; yet the public health implications of these emerging dynamics are multi-dimensional and remain under-researched. They reach beyond the potential to increase demand for health care in destination communities, and raise questions about healthy environments, cultures of health care (that is, Western versus traditional), and discourses of inclusion, exclusion and blame.

This volume

This book represents one of the first efforts to collect into a single volume the work of a growing number of researchers who simultaneously study migration and health in the context of Asia. It brings together issues and debates related to the intersection of health and migration. It includes empirical case studies from countries at various levels of socio-economic development, ranging from 'transition' countries to economically highly developed countries, with studies in Bangladesh, China, Japan, India, Indonesia, the Philippines, Singapore and Laos. The chapters not only define health in a variety of ways, using measures of physical health and functional health (mobility and disability), but also examine the health-related behaviours, social suffering and surveillance of migrants, as well as relative health outcomes for migrants and non-migrants. Later chapters, in their consideration of the health of those left behind, shift the focus from the migrants themselves to their communities of origin. The diverse geographies of migration are recognized in the discussion of transnational, national, rural to urban, and upland to lowland migratory movements. The country studies adopt different methodologies, with five chapters (Hugo; Jatrana and Chan; Feranil; Kuhn; and Roy and Nangia) being primarily quantitative in approach, while others report data from qualitative surveys.

Overall, the multi-disciplinarity of this research field is exposed, with authors drawing on concepts and methods from several academic disciplines including anthropology, demography, sociology, human geography and public health. The chapters, when taken together, constitute the first serious attempt to provide an insight into the potential scope of research on health and migration in an Asian context, and are thus an important benchmark for future work in this field.

Key themes

At the core of each study included in this volume is a broader question of how we can properly understand the health vulnerabilities of migrants, and communities of origin and destination. The findings vary considerably, depending on the specific characteristics of migrants, type of migration (regular or irregular, national or transnational, rural to urban, or highland to lowland), the particular health measures used, and the theoretical and methodological approach taken (for example, a behavioural versus an institutional approach). Although none of the studies in this volume include all of the important variables associated with migration and health, each makes a serious attempt to investigate specific linkages, and to reach conclusions that go beyond the simplistic assertion that migration has positive or negative consequences for the health of migrants, host communities or families left behind. Nine key themes structure the discussion.

1 Migrants, sexual behaviour and vulnerability to STDs/HIV/AIDS

Although the rapid spread and high prevalence of HIV/AIDS in some areas of Asia has been invariably linked to migration, we are still far from understanding in detail just how, and to what extent, migration affects the spread of HIV. There remain large gaps in our knowledge of the role of migration in explanations of why levels of infection in one region are higher than in another region. There is also a need for the conceptual refocusing of research on the social and sexual disruption that accompanies migration and mobility. It is in this context that Hugo (Chapter 2) examines possible links between the current incidence and spread of HIV in Indonesia, and population mobility, insofar as it can be established given the dearth of appropriate data. Hugo also discusses the extent to which population mobility is likely to influence the spread of HIV infection in the future.

Hugo's findings suggest that high levels of population mobility are not neatly correlated with high levels of HIV infection. Rather, the latter are influenced by the type of movement, the context in which it occurs and the behaviour of the movers themselves. Higher risk of HIV infection appears to be associated with specific types of mobility, although the research evidence is as yet limited. These include temporary worker movement to isolated work sites, such as mining, construction, plantation and sawmilling settlements; rural to urban migration; circulation involved in some types of work, such as transport, fishing and seafaring; movement associated with the large internal displacement of population due to conflict; and some international labour migration. Hugo anticipates that the nexus between the commercial sex industry, concentrations of migrant workers and HIV infection is a key to the future course of transmission of the infection in Indonesia and urgently requires further research. Hence, he emphasizes the need to focus more behavioural sentinel surveillance activity and programme inventory on mobile populations of various types.

2 Modernization and health among ethnic groups

Understandings of migration and health must resist treating migrants as a homogeneous group, and a more nuanced approach requires a recognition of diversity in the social contexts of migration. In Chapter 3, Lyttleton considers the relocation of an ethnic minority group from the highlands to the lowlands in Asia's Golden Triangle as an important form of migration that crosses a significant boundary, not defined by distance or national borders but by elevation. Although mobility has been commonplace in the highlands of the Upper Mekong for centuries, contemporary migration downhill, even over short distances, requires significant social and cultural change and adaptation that can be as disruptive as transnational movement. Lyttleton focuses on the Akha of Northwest Laos who are moving out of subsistence life-styles in the mountains and entering new forms of social arrangement that require new social competencies – characterized most strongly by sedentary life-style, wage labour and cash economy. Voluntary and involuntary relocation has created a crucible where lowland Lao, Chinese labourers and resettled Akha engage in reconfigured social relations with specific health consequences.

Even though moving down from the mountains improves access to state medical services and market resources, the negative implications for health remain pronounced. Lyttleton draws attention to the health impacts implicit in highland–lowland migration from the perspective of social suffering. Here the intention is to consider health in ways beyond the presence or absence of disease, by looking at how the new social relations which emerge within processes of nationalist assimilation incur forms of marginalization and exclusion. In Northwest Laos, the uneasy intersection of different cultural systems promotes inequitable economic and social interactions that foster social suffering.

While empirical measurement of the health impacts may not always be possible, Lyttleton argues that the effects of social suffering are anything but ephemeral. Importantly, this perspective draws attention to social relations, rather than to material conditions, as the underlying preconditions for ill-health. Lyttleton demonstrates how, despite greater access to material benefits associated with modernization, social suffering among the Akha creates increased vulnerability to a range of health threats, including heightened vulnerability to HIV infection, changing forms of drug abuse and embodied forms of psychic malaise. These health threats emerge as villagers experience rapidly changing social values, socio-economic stratification and ethnic exploitation as part of greater involvement in a cash economy.

3 Migration, health and the state

The greater vulnerability of migrants is also played out at the national scale. It has been evident in many circumstances that migrant workers are frequently seen as a mobile, floating and even 'disposable' population, making them vulnerable to repatriation, neglect and discriminatory policies founded on classed constructions of ethnicity and nationality. The onslaught of SARS in Singapore and elsewhere provides an opportunity to examine the intersection of national concerns over the 'behaviour' of migrant workers, and over the construction of health as a problem of national security. The SARS crisis, played out over a period from March to July 2003, sent many countries into shock, forcing governments to respond with immediacy. It is during a crisis such as this that migrant workers are exposed as mobile, transient and not always controllable, requiring the implementation of quarantine measures and other solutions to shut out the killer virus, and to prevent it from crossing national borders.

Wong and Yeoh (Chapter 4) focus on Singapore and examine how state policy controls migrant workers, and how public discourse contributes to the social construction of migrants as 'health threats' during a time of crisis (SARS). They trace the main threads of government rationalization over the treatment of migrants seen as necessary for curtailing the spread of SARS, while at the same time negotiating the delicate balance needed to maintain a healthy, active foreign worker pool for the sake of the economy.

Wong and Yeoh argue that the government's attitudes, some of which were built upon discourses that emerged during the time of economic boom and global restructuring, were often contradictory and did not always reflect the patterns of the spread of SARS. First, migrants underwent a higher degree of surveillance than did Singaporeans, in spite of Singaporeans comprising the vast majority of SARS victims and 'infectors'. Second, although low-skilled labour migrants have always been more tightly controlled than migrant workers in the professional and managerial categories, it was low-skilled foreign workers who faced harsher penalties and a higher level of checks than workers at the other end of the socio-economic spectrum, even though SARS appears to have crossed into Singapore through the movement of middle-class Singaporean travellers. Wong and Yeoh show that even where the cause–effect relationship between migration and health is extremely tenuous, health security concerns become a ready platform for expressing deep-rooted prejudices against migrant workers in public discourse, as well as a *raison* d'être for strengthening the role of the state as gatekeeper of the nation through its finely tuned, selective control of national borders.

4 Do migrants face greater health risks than non-migrants?

The multiple vulnerabilities of different migrant groups within different local and national settings raise questions of how their health compares with that of non-migrant groups. Answering such questions is difficult, however, since it involves addressing several key issues, including how migrants' health status differs from the health status of the native-born, and how the health status of migrants evolves over time after their arrival in the host region or country. In what ways and how quickly, for example, does the health status of migrant populations change through exposure to the social, cultural, economic and physical environment within the host area? Which covariates are associated with health in general, and migrants' health in particular? Thus far there has been little research into these issues in Asia.

There are indications in literature on Western settings that some migrant populations may experience important health disparities, relative both to other migrant groups and native-born populations (Newbold and Danforth 2003; Sundquist and Johansson 1997). Indeed, a critical review of the literature demonstrates that although migrants are generally healthy at the time of arrival (the 'healthy migration effect'), their health status and that of the next generation tends to decline and converge (downwards) towards that of the native-born population. But why does the health status of some migrants deteriorate with duration of residence? Some researchers attribute the deterioration to the uptake of unhealthy life-styles, including poor dietary habits, smoking, and/or drinking, upon arrival in the host country (Frisbie *et al.* 2001).

However, it may take decades for the negative health implications of behavioural change to become apparent and other, perhaps countervailing, factors may intervene in the shorter term. Any overall decline in the health status of migrants thus seems likely to be the result of a complexity of factors operating within the broad context of the migrants' social, political, economic and cultural position within the host community. The local determinants of population health may be magnified within a migrant population as loss of socio-economic status, social networks, discrimination, language barriers, and lack of knowledge about the availability of health facilities contribute to declines in health (Newbold 2005). Moreover, unfamiliarity with the health care system may lead to relative underuse of preventive health screening and underdiagnosis and treatment of health problems, which in turn may lead to a worsening of health status over time. In empirical studies, however, tracking health status longitudinally is problematic, especially where health status is self-assessed. In the Canadian context, McDonald and Kennedy (2004) find that immigrants' use of basic components of health care approaches native-born levels. Nevertheless, improved access and use of health services can lead to increased recognition and reporting of conditions, and consequently poorer self-assessed health. Understanding the trajectory of change in migrants' health status thus requires close attention to the health measures employed.

In Chapter 5, Jatrana and Chan examine the population of Singapore in a quantitative analysis of the influence of country of birth (born in Singapore versus born in China, Malaysia or elsewhere) on self-reported functional disability among older adults. The question they explore is whether nativity is an independent risk factor for reporting a functional disability, or whether nativity differences in self-reported functional disability merely reflect differences in socio-economic and demographic characteristics. This is an important question for policy makers. If nativity differences in self-reported functional disability can be attributed to demographic and socio-economic characteristics, rather than to nativity itself, then functional disability reduction programmes can be targeted to compensate for the differences between groups based on traits such as education, employment and income. If, however, nativity itself is a significant determinant of functional health, then a culturally based barrier may exist, in which case simply expanding the availability of health services or altering socio-economic conditions of people may not enhance functional health.

Jatrana and Chan's results indicate that nativity is not a significant risk factor for reporting functional disability once demographic factors are controlled for. Thus the difference in functional disability between native- and foreign-born older adults is largely explained by the demographic variables in their model. Socio-economic, health-related behaviour and social networking factors do not affect the nativity differences in functional disability to any significant degree. Rather, Jatrana and Chan attribute the difference to the erosion of the healthy migrant effect over time and the ageing of the foreign-born population who are much older in their demographic composition, compared to the native-born population.

Feranil (Chapter 6) also explores the differences in health between migrant and non-migrant populations, focusing on internal migrants and using a more specific measure of health. He examines maternal anaemia among migrants and non-migrants in his analysis of a range of data from a survey conducted in disadvantaged communities in one of the island groups in the Philippines. His findings indicate that migrants tend to be healthier than the native-born, supporting the argument that migration is a selective process, favouring healthier individuals. Although other factors such as age, education, number of births, living conditions, and the mother's previous practice of taking iron supplements also influence maternal anaemia, he stresses that the health of both the native-born and migrants are equally important, particularly in disadvantaged areas. Feranil recommends that public health services and nutrition campaigns should address the predicament of both groups (migrants as well as non-migrants), a recommendation endorsed by Jatrana and Chan (Chapter 5).

5 *Irregular/unauthorized migration and health*

The importance of the social and political context of migration is again emphasized in Chapter 7, where Asis draws attention to the ways in which migrants' unauthorized status in receiving societies can affect their health. As documented in various settings, unauthorized migrants are generally more vulnerable to health risks than are legal migrants. The working and living conditions of unauthorized Filipino migrants in Sabah are no different. The hostile attitude of the general society towards Filipinos in Sabah, the precarious terms of their employment, the lack of services provided by the Philippine government to its nationals in Sabah (a situation arising from the unresolved Philippine claim on Sabah), and the absence of non-government organizations – an alternative source of assistance for migrants in other receiving societies – contribute to the vulnerability of irregular Filipino migrants in Sabah. In this context, it is not surprising that health concerns do not figure as much in migrants' assessment of their situation as do the uncertainties of their unauthorized status. Their lack of legal status and associated rights circumscribe the options available to migrants, including options for health care.

Asis finds that when migrants fall ill, they do not access government health facilities; rather, they seek out private practitioners, opt for traditional healers, or return to the Philippines to seek medical help. In other words, the concern of local residents that migrants put strain on local medical facilities seems to be more perceived than real. In clarifying the linkages between irregular migration and health, Asis suggests the need to consider other variables that could explain the varying degrees of health vulnerabilities and access to health care of unauthorized migrants. Migrants' ethnic background, whether or not they migrated with family members, and their occupation, are some of these intervening variables, suggesting diversity, as well as commonality, in the circumstances of this migrant group.

6 *Migration, differential access to health and civil society*

In Asia, immigration laws typically permit unskilled migrants to work for a short period of time while ignoring their rights as residents and workers. Official neglect of human rights for migrant workers has created serious problems for them and for their host society, since they are regularly exposed to labour accidents and injuries, abuse, violence, illness, depression, and epidemics such as HIV/AIDS. Nonetheless, citizens of most Asian labour importing countries remain unaware of these problems, and therefore pay little attention to the plight of foreign labourers. As a result, throughout Asia, non-governmental organizations (NGOs) and civil groups are the only agents dedicated to the protection of migrants' rights. In response to the critical importance of personal and public well-being, Asian NGOs have commonly worked to increase public awareness of health issues, while providing health services and health training for migrant men and women.

The case of Japan discussed by Yamanaka (Chapter 8) indicates the importance of international human rights laws as a framework for encouraging governments to eliminate discrimination against unskilled migrants embedded in law, bureaucracy and conventional social practices. More research will be necessary if we are to understand how these two forces – international law applied from 'above' and grass-roots activism emerging from 'below' – can bring about significant change to governmental policies and public attitudes towards migrants' rights, including their access to health and medical care. These rights are fundamental to the public health and social welfare of the nation at large.

The role of civil society is important and indispensable not only for the welfare of transnational migrants but also, in the case of China, for internal migrants whose move from a rural to an urban area is not 'authorized' by the government. Xiang (Chapter 9) highlights this in his analysis of health risks associated with rural–urban migration in China. These migrants, the so-called 'floating population', face special health problems not only because they are a new social group, but more importantly because they cannot be incorporated into the state system, yet at the same time the state remains the sole provider of social welfare. As the volume of rural–urban migration increases, this bureaucratic exclusion will have a growing impact on public health in China's cities. Xiang concludes that the Chinese government must recognize that it is unrealistic to provide for all its citizens' welfare needs without cooperating with other social forces, and that an active civil society may thus be indispensable to maintaining social stability within an increasingly diversified population.

7 *An institutional approach to understanding migration and health*

In linking migration and health, particularly in studies of migration and HIV, most of the existing literature adopts a 'behavioural approach', which assumes that certain migrant behaviour patterns are disease-prone (Brocker-hoff and Biddlecom 1999; Caldwell *et al.* 1997; Hunt 1989; Jochelson *et al.* 1991; for a literature review see Yang 2004). Moving away from this perspective, Xiang (Chapter 9) proposes an 'institutional approach' which holds that migrants face particular health problems because of their position in the established social system. Based on his fieldwork and documentary research in China, Xiang demonstrates that migrants' health problems are fundamentally the result of existing formal institutional arrangements. The fact that migrants face high health risks at work is seemingly a result of market segmentation, but fundamentally it is attributable to the unbalanced relationship between migrants, enterprises and local government, which is, in turn, related to the *hukou* system and the current economic growth regime in China. The existing institutional arrangements not only render migrants vulnerable, but also impede them from being included in the formal medical care system.

More specifically, Xiang identifies the major reasons for this exclusion as including various vested interests and the government's immediate goal of reforming the social security system; the gap between the medical care system in urban and rural areas, which makes it impractical to provide migrants with formal medical care in the cities; the localized, and therefore geographically fragmented medical provision, which discourages migrants from joining the system; and finally the informal employment relationship prevalent among migrant workers, which conflicts with the medical care system's reliance on formal employment contracts for implementation.

In terms of policy recommendations, the behavioural approach tends to focus on policies that directly affect migrants' health-related behaviours, while the institutional approach draws attention to the fact that medical policies are embedded in various other institutional structures and, accordingly, stresses the importance of examining underlying institutional arrangements in order to identify realistic strategies for policy change. The focus on structural determinants of risk emphasized by the institutional approach provides a useful corrective to the dominant behavioural paradigm which encourages the ascription of blame to individuals for indulging in risky behaviour. Nevertheless, it offers only a partial, albeit valuable window on to the complex interrelationships between migration and health in Asia. As Craddock (2000) argues in her discussion of the geography of AIDS, we need to adopt a combination of approaches in order to develop a more effective theoretical framework for evaluating vulnerability to ill-health. In the case of Asia, there is an urgent need both for more empirical research to establish the scope of diversity across migrant groups and for greater attention to the ways in which we conceptualize migrants' vulnerability.

8 *Migration and the health of left-behind populations*

All the chapters outlined above are devoted to understanding, in various ways, the health of migrants themselves. In the wider literature, while relatively few studies have been carried out on the topic of migrant health in general, there are even fewer studies of the impact of migration on the health of those who are left behind. Recent theoretical and methodological advances in the study of migration have facilitated analysis of migration's impact not merely on the health of migrants, but also on the health and well-being of family members left behind in the migrant's household and community of origin. A few recent papers have examined the impact of parents' migration on the health of children left behind (Frank and Hummer 2002; Kanaiaupuni and Donato 1999), but none has yet looked at the impact of adult children's migration on the health of their older parents. Kuhn (Chapter 10) addresses this crucial intersection between global concern over population ageing and migrant social networks by modelling the impact of adult children's migration on the health and survival of a cohort of respondents age fifty and above living in Matlab, a rural area of Bangladesh where high rates of out-migration and remittance receipt make migration a key aspect of social and economic life.

By linking detailed household and health survey data to event records from an ongoing Health and Demographic Surveillance System (HDSS), Kuhn shows that, far from having a deleterious impact on elders, adult children's migration has a strongly positive effect on health and survival. Health is improved largely through the effects of the migration of sons in this traditionally patrilineal society, with a significant impact in relation to both internal and international migration. While migrant educational attainment can explain the effects of internal migration, the effects of international migration cannot be explained by education or by measures of financial transfers in the year of the survey. The failure to find any relationship between remittances and health calls into question the value of transfers as an indicator of the *long-term* strength of support from migrant children, and raises questions over the difficulties of relating remittance and health data at the individual level. Kuhn concludes that the international migration of sons is conducive to parents' health, although the mechanism is still unclear. In Matlab, financial transfers apparently have little effect on the health of older parents.

However, the negative impacts of migration on those left behind can be seen clearly in studies on migration and the spread of STDs and HIV/AIDS. A study in Delhi, India found that migrants living without their families (or those whose families are left behind) are more likely to engage in paid sex and extra-marital sexual relationships (Mishra 2004). In this volume, Roy and Nangia (Chapter 11) use evidence from rural Bihar, India to show that wives left behind by migrant men report a significantly higher prevalence of reproductive morbidity than do wives of non-migrant men. Results from this study also indicate that determinants of reproductive morbidity differ between the left-behind wives and the wives of non-migrants. Modelling results show that for the wives of non-migrant men, extra-marital relations significantly affect reproductive morbidity. In contrast, for the wives left behind by migrant men, a lower standard of living, shorter marital duration, the migration of the husband to Western states, the timing of migration, and the husband's life-style (habits such as gambling, substance abuse, chewing tobacco and so on) are significant contributors to reproductive morbidity. Possibly, in the absence of their husbands, wives who are left behind are unable to take care of their reproductive morbidity, particularly when they feel shy about discussing it with anyone, including a doctor. Overall, the reproductive health problems of women are neglected as marital duration increases, as well as when husbands adopt life-styles that might lead to extra-marital relations or divert financial resources away from necessary health care.

Both Kuhn, as well as Roy and Nangia, tackle important questions about the impact of out-migration on population health in communities of origin. Both studies focus on South Asia where traditional patriarchal values prevail. Further research in other cultural contexts could usefully explore the different vulnerabilities of those left behind in diverse settings across Asia. This evidence base is necessary for the design of effective policies to mitigate the negative health impacts of out-migration on communities of origin.

9 *Migration and health: a policy perspective*

Migration has always posed particular problems for economic and social planning. The same is true for health planning. Governments are under pressure to give priority to the 'permanent' population (that is, long-term residents *vis-à-vis* temporary migrants in the case of internal migration; citizens *vis-à-vis* labour migrants in the case of transnational migration). However, in relation to labour migrants, there is frequently a willingness on the part of the governments of both sending and receiving countries to trade off migrant welfare against the economic benefits that a largely unregulated and unprotected flow of labour migrants brings. We have only begun to identify the dimensions of vulnerability of migrants and left-behind families but, as a first step to addressing some of the problems and risks discussed in the chapters of this book, governments across Asia should be prevailed upon to fully implement the policy guidelines developed by the International Labour Organization in the 1990 International Convention on Migrants' Rights.

Assuming that Asian governments have the political will to look after the interests of migrants, including their health needs, Jones (Chapter 12) outlines the different levels at which policy may be directed. At the 'coal-face' there is the need to deal with the immediate health needs of migrants. At one remove, steps are needed to ensure that the context in which migrants find themselves is improved, through such means as ensuring that employers comply with regulations and provide satisfactory working conditions, along with checks on conditions in employer-supplied accommodation. At a further remove there is the need to improve understanding of the political, social, economic and cultural contexts of those migratory movements that have come to be associated with particular health problems. In the policy arena, civil society actors have an important role to play and may be more effective than governments in addressing the health needs of migrants and their families.

Effective policy requires better understanding of the issues and the will to do something about them. On the former, more targeted research is required, and improved assessment of the implications of research already conducted. On the latter, the needs of both migrants and left-behind families require a voice that can influence policy making, perhaps through NGO and other non-governmental groups. As new forms of transnational and internal migration emerge in Asia, studies of the many intersections between migration and health have a valuable contribution to make in shaping public health agendas. It is hoped that the contribution of this book will be to encourage further research into migration and health in Asia.

References

Anson, J. (2004) 'The migrant mortality advantage: a 10 month follow-up of the Brussels population', *European Journal of Population*, 20: 191–218.

Berry, J.W. (1997) 'Immigration, acculturation and adaptation', *Applied Psychology: An International Review*, 46: 5–34.

Bollini, P. (1992) 'Health policies for immigrant populations in the 1990s: a comparative study in seven receiving countries', *International Migration*, 30: 103–19.

Borman, E. (2004) 'Health tourism – Where healthcare, ethics, and the state collide', *British Medical Journal*, 328 (7431): 60–1.

Boyle, P.J. (2004) 'Population geography: migration and inequalities in mortality and morbidity', *Progress in Human Geography*, 28, 6: 767–76.

Boyle, P.J. and Graham, E. (2003) 'Understanding the geographies of migration and health', Paper presented in an International Workshop on Migration and Health in Asia, 22–24 September, Bintan, Indonesia.

Boyle, P.J., Halfacree, K.H. and Robinson, V. (1998) *Exploring Contemporary Migration*. London: Longman.

Brockerhoff, M. and Biddlecom, A. (1999) 'Migration, sexual behavior and the risk of HIV in Kenya', *International Migration Review*, 33, 4: 833–56.

Caldwell, J.C., Anarfi, J.K. and Caldwell, P. (1997) 'Mobility, migration, sex, STDs, and AIDS: an essay on Sub-Saharan Africa with other parallels', in G. Herdt (ed.) *Sexual Cultures and Migration in the Era of AIDS: Anthropological and Demographic Perspectives*. New York: Oxford University Press, pp. 41–54.

Chung, R.C.Y. and Kagawa-Singer, M. (1993) 'Predictors of psychological distress among Southeast Asian refugees', *Social Science and Medicine*, 36: 631–9.

Cookson, S.T., Carballo, M., Nolan, C.M., Keystone, J.S. and Jong, E.C. (2001) 'Migrating populations – a closer view of who, why and so what', *CDC (Centre for Disease Control and Prevention) Newsletter*, 7, 3: 1–3.

Craddock, S. (2000) 'Disease, social identity, and risk: rethinking the geography of AIDS', *Transactions of the Institute of British Geographers*, 25: 153–68.

Evans, J. (1987) Introduction: migration and health', *International Migration Review*, 21, 3: v–xiv.

Findley, S.E. (1988) 'The directionality and age selectivity of the health–migration relation: evidence from sequences of disability and mobility in the United States', *International Migration Review*, 22: 4–29.

Fitinger, L. and Schwartz, D. (eds) (1981) *Strangers in the World*. Bern: Hans Huber.

Frank, R. and Hummer, R.A. (2002) 'The other side of the paradox: the risk of low birth weight among infants of migrant and nonmigrant households within Mexico', *International Migration Review*, 36, 3: 746–65.

Frisbie, W.P., Youngtae, C. and Hummer, R.A. (2001) 'Immigration and the health of Asian and Pacific Islander adults in the United States', *American Journal of Epidemiology*, 153, 4: 372–80.

Harding, S. and Balarajan, R. (2001) 'Longitudinal study of socio-economic differences in mortality among South Asian and West Indians', *Ethnicity and Health*, 6, 2: 121–8.

Hull, D. (1979) 'Migration, adaptation, and illness: a review', *Social Science and Medicine*, 13A: 25–36.

Hunt, C.W (1989) 'Migrant labor and sexually transmitted diseases: AIDS in Africa', *Journal of Health and Social Behavior*, 30: 353–73.

Jochelson, K., Mothibeli, M. and Leger, J. (1991) 'Human immunodeficiency virus and migrant labor in South Africa', *International Journal of Health Services*, 21, 1: 157–73.

Jones, E.E. and Korchin, S.J. (eds) (1982) *Minority Mental Health*. New York: Praeger.

Kanaiaupuni, S.M. and Donato, K.M. (1999) 'Migration and mortality: the effects of migration on infant survival in Mexico', *Demography*, 36, 3: 339–53.

Kaplan, M. and Marks, G. (1990) 'Adverse effects of acculturation: psychological distress among Mexican-American young adults', *Social Science and Medicine,* 31: 1313–19.

Kuo, W.H. and Tsai, Y. (1986) 'Social networking, hardiness and immigrants' mental health', *Journal of Health and Social Behaviour,* 27: 133–49.

Marmot, M.G. (1989) 'General approaches to migrant studies: the relation between disease, social class and ethnic origin', in J.K. Cruickshank and D.G. Beevers (eds) *Ethnic Factors in Health and Disease.* London: Wright, pp. 12–17.

McDonald, J.T. and Kennedy, S. (2004) 'Insights into the "healthy immigrant effect": health status and health service use of immigrants to Canada', *Social Science and Medicine*, 59: 1613–27.

Mishra, A. (2004) 'Risk of sexually-transmitted infections among migrant men: findings from a survey in Delhi', *Asian and Pacific Migration Journal*, 13, 1: 89–106.

Morales, L.S., Lara, M., Kington, R.S., Valdez, R.O. and Escarce, J.J. (2002) 'Socioeconomic, cultural, and behavioral factors affecting Hispanic health outcomes', *Journal of Health Care for the Poor and Underserved*, 13, 4: 477–503.

Newbold, K.B. (2005) 'Self-rated health within the Canadian immigrant population: risk and the healthy immigrant effect', *Social Science and Medicine*, 60: 1359–70.

Newbold, K.B. and Danforth, J. (2003) 'Health status and Canada's immigrant population', *Social Science and Medicine*, 57: 1981–95.

Norman, P., Boyle, P.J. and Rees, P.H. (2005) 'Selective migration, health and deprivation: a longitudinal analysis', *Social Science and Medicine*, 60: 2755–71.

Rashid, S.R. (2002) 'The right of health of migrant workers', *Udbastu, Newsletter on Refugee and Migratory Movements*, 21: 3–5.

Rosenwaike, I. (1990) 'Mortality among three Puerto Rican populations: residents of Puerto Rico and migrants in New York City and in the balance of the United States, 1979–1981', *International Migration Review*, 24: 684–702.

Shuval, J. (1993) 'Migration and stress', in L. Goldberger and S. Breznitz (eds) *Handbook of Stress: Theoretical and Clinical Aspects* (2nd edn). New York: The Free Press, pp. 677–91.

Srithanaviboonchai, K., Choi, K.H., van Griensven, F., Hudes, E.S., Visaruratana, S. and Mandel, J.S. (2002) 'HIV-1 in ethnic Shan migrant workers in northern Thailand', *Aids*, 16, 6: 929–31.

Sundquist, J. (1994) 'Refugees, labour migrants and psychological distress', *Social Psychiatric Epidemiology*, 29: 20–4.

Sundquist, J. and Johansson, S.E. (1997) 'Long-term illness among indigenous and foreign-born people in Sweden', *Social Science and Medicine,* 44: 189–98.

Taran, P.A. (2002) 'Migration, health and human rights', *Migration and Health,* International Organisation for Migration (IOM), 2/2002: 1–4.

Yang Xiushi (2004) 'Temporary migration and the spread of STDs/HIV in China: is there a link?', *International Migration Review*, 38, 1: 212–35.

2 Population movement in Indonesia

Implications for the potential spread of HIV/AIDS

Graeme Hugo

Introduction

The most recent estimates suggest that between 90,000 and 130,000 Indonesians were infected with HIV in 2002 – an infection rate of 0.1 per cent (Ministry of Health, Republic of Indonesia 2002). These rates are much lower not only than many sub-Saharan countries, but also than other Southeast Asian countries such as Cambodia, Thailand and Myanmar which have serious HIV/AIDS epidemics (Brown 2002). However, there is little room for complacency in the world's fourth largest nation because, first, in recent years there have been substantial increases in the overall number of reported cases of both HIV and AIDS, as well as evidence of increased infection in some samples of high vulnerability groups (Hugo 2001). In 2002 it was estimated that 80,000 people would be infected by HIV in 2003 (Ministry of Health, Republic of Indonesia 2002: 5). Second, several commentators see Indonesia as being vulnerable to the possibility of a substantial increase in HIV transmission due to the presence of conditions which have elsewhere been associated with the spread of the disease (Kaldor 2000).

Recent estimates also suggest that the number of people vulnerable to HIV infection in Indonesia is estimated at between thirteen and twenty million (Ministry of Health, Republic of Indonesia 2002). Since HIV is spread by actions requiring close contact between individuals – sexual intercourse, transferral of contaminated blood and the sharing of contaminated hypodermic needles[1] – understanding the *mobility* of infected individuals and groups at high risk of becoming infected is crucial to understanding the patterns of transmission and spread of the disease, and for developing interventions to slow down or stop its spread. It is therefore important to consider the potential role of population mobility in the spread of HIV/AIDS. Population movement has been established as an important independent risk factor for HIV infection in many contexts (see e.g. Appleyard and Wilson 1998; Decosas *et al.* 1995; Gardner and Blackburn 1996: 10; Haour-Knipe and Rector 1996; Herdt 1997), building on a substantial literature which has established the links between population movement and the spread of infectious disease through history (see e.g. Prothero 1965, 1977; Wilson

1995). In Indonesian work on HIV/AIDS this issue has received little or no attention, but I argue that the high levels of population mobility in Indonesia, the nature of that mobility, the characteristics of the movers and the circumstances of their movement have important potential implications for the rapid diffusion of HIV infection in Indonesia.

Unlike the case elsewhere in Asia, there is little explicit research which has explored the links between HIV/AIDS and population mobility in Indonesia (Walzholz 2003). I seek to draw together knowledge of population mobility in Indonesia, and assess its likely implications for the spread of HIV/AIDS and for planning interventions which seek to prevent the spread of the disease. I argue that contemporary population mobility in Indonesia has the potential to facilitate the rapid spread of the disease, as it has done in parts of Africa (Brockerhoff and Biddlecom 1999; Caldwell *et al.* 1997; White *et al.* 2001). Hsu and Du Guerny (2000: 2) point out:

> In Africa, the epidemic was for many years considered to be urban-based which [led] to an ignoring of the vulnerabilities of rural populations, returning migrants and other mobile groups such as travelling salesmen. This neglect of the rural populations led to the epidemic spreading unnoticed for a long period.

Such a neglect of the role of population mobility in the early stages of the epidemic should not be allowed to occur in Indonesia or elsewhere.

Conceptualizing the link between population mobility and HIV/AIDS

One of the most universal features of migration is that migrants are frequently stigmatized by majority communities at destinations. They become scapegoats and are blamed often, if not usually unfairly, for perceived negative aspects of destination life including crime, disease, social conflict and the lowering of work conditions. Hence it is important for movers *not* to be seen to *automatically* be at higher risk of HIV infection than non-migrants. Indeed, there are many cases where migrants have a lower incidence of infection than do non-migrants.

Among the large Indonesian migrant worker population in Malaysia there has long been a concern that they may be associated with the spread of infectious disease. This is often part of a general 'scapegoating' of the migrant workers which sees them as the cause of a range of social, health and economic problems in Malaysia. Indeed, in 1992 it was reported in the Malaysian press (*The Nation*, 31 July 1992) that 30 per cent of those who registered for work permits in Malaysia 'were carrying the AIDS virus'. The great majority of people applying for such permits were Indonesians. However, another newspaper (*Straits Times*, 6 August 1992) reported the Malaysian Minister of Health as saying that a random sample of 5,000 foreign workers who were tested for HIV found only twelve HIV positive cases.

Skeldon (2000: 1–2) identifies three critical issues in the population mobility/HIV/AIDS relationship. First, it is not so much migration that is important as the *behaviour* of migrants. It is the combination of migration and high-risk behaviour (unprotected sexual intercourse or sharing needles by injecting drug users) that is central. Second, the most 'at-risk' group are not so much migrants as conventionally defined but non-permanent movers. Finally, mobility may place people in high-risk situations. Thus the relationship between population mobility and the spread of HIV/AIDS is clearly not a simple one. Movers are only at higher risk of HIV infection than non-movers if they engage in high-risk behaviour to a greater extent than do non-movers. Clearly this is the case for *some groups of movers*, but not all. Nevertheless, having become infected, they can then transfer the disease to other places to which they move and to their home areas if they return. Some movers will be more at risk than non-movers at origin and destination because they are at greater risk of engaging in risk-taking behaviour. This is, first, because the selectivity of the migration process means that people with higher risk are more mobile, and, second, because the effects of the context of the migrants, especially at the destination, may predispose them towards high-risk behaviour.

Several writers have developed conceptual models which make these relationships explicit. Most notably, Brockeroff and Biddlecom (1999: 834) show that the sexual behaviour of migrants is a function of their characteristics (gender, marital status, education level, socialization, ethnicity, religion and age), the process of migration (separation from spouse/partner), and the new social environment at the destination (permissiveness, support network, income earning opportunities and so on). These influence their perception of the risks and consequences of their behaviour. The authors use the 1993 Demographic and Health Survey for Kenya to test their model, and conclude that while there was evidence of a possible relationship between migration and high-risk sexual behaviour in Kenya, the relationship was not consistent across different gender and migration flow groups.

Yang (2002a, 2002b) and Smith and Yang (2002) argue that it is important in examining the role of migration in the spread of HIV to go beyond migration's role as a 'virus carrier and population mixer' and to identify and understand the underlying mechanisms which make migrants more susceptible to risk-taking behaviour. They see migrants as being more vulnerable because, first, they are more socially and economically marginalized and spatially isolated at the destination, and, second, because of diminished social control at the destination and their removal from much of the influence of community norms due to migration. However, this is not always the case, since much research in Asia shows that many migrant workers move along well-established migration routes and stay with relatives and friends from their origin areas when at the destination (Hugo 1978).

HIV/AIDS in Indonesia

In Indonesia the reported numbers of AIDS and HIV positive cases, as Table 2.1 shows, are quite small, although there has been a substantial increase in numbers in recent years. There is little room for complacency because reported cases are only a fraction of actual cases. The UN, for example, has estimated that in China, in a similar situation, only 5 per cent of all HIV/AIDS cases are reported (Bickers and Crispin 2000: 38). There has also been a rapid increase in reported cases in the past few years, and the conditions in Indonesia are seen to be favourable to a rapid spread of the disease, perhaps in a similar way as to what has occurred in Thailand (Kaldor 2000). One estimate puts the number of Indonesians living with HIV in 2001 at between 80,000 and 120,000 (Directorate General of Communicable Disease Control and Environmental Health 2001: 4). The same report points out that 'the consistently lower rates of HIV infection recorded over the last decade are a thing of the past'.

The spread of HIV/AIDS in Indonesia appears to have gone through three stages (Utomo, personal communication, February 2001) after its initial introduction in 1987. In the early years it was confined to small groups and was spread mainly by homosexual contact. By the 1990s the main method of transmission was through heterosexual contact. Since 2000 there has been a major increase in the amount of transmission via shared needles by intravenous drug users, although heterosexual sexual transmission remains the main mode of transmission.

Table 2.1 Number of HIV/AIDS positive cases in Indonesia 1987 to November 2000

Year	AIDS	HIV (+)	Total
1987	2	4	6
1988	2	5	7
1989	3	4	7
1990	5	4	9
1991	12	6	18
1992	10	18	28
1993	17	96	113
1994	16	71	87
1995	20	69	89
1996	32	105	137
1997	34	83	117
1998	74	126	200
1999	47	178	225
2000 (up to Nov)	172	344	516
Cumulative total	446	1,113	1,559

Source: UNAIDS (Jakarta) (2001).

In the present context it is important to point out that there are important variations in the incidence of HIV/AIDS between different parts of Indonesia. Figure 2.1 shows the numbers of reported HIV positive infections, AIDS cases and deaths from HIV/AIDS in each of Indonesia's provinces as of the end of November 2000 per 100,000 residents. It is immediately apparent that HIV/AIDS reported cases are most prevalent in a few provinces. Importantly, in each of the areas in which HIV notification rates are greater than the national average, there is a high level of population mobility.

In parts of West Papua there is an HIV/AIDS epidemic well underway. The first HIV/AIDS cases were reported in Merauke in 1992 – six cases. By March 2001 there were HIV/AIDS cases in eleven of the thirteen *kabupaten* in the province and a total of 546 reported cases (Indonesian Directorate of Direct Transmitted Disease Control 2001). This represented a notification rate of six per 100,000 – twenty-eight times higher than the national rate (0.22 per cent). Merauke is the only place in Indonesia where the notification rate is more than 1 per cent of the population. The bulk of people with the disease are indigenous West Papuans, among whom there are a number of local behavioural and cultural elements which favour the spread of HIV/AIDS, especially having multiple partners. Table 2.2 shows the incidence of HIV in particular high-risk groups in West Papua in 2000. The

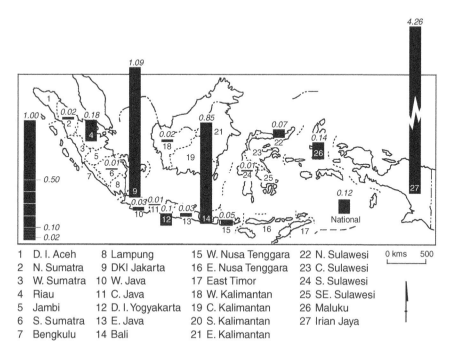

1	D. I. Aceh	8	Lampung	15	W. Nusa Tenggara	22	N. Sulawesi
2	N. Sumatra	9	DKI Jakarta	16	E. Nusa Tenggara	23	C. Sulawesi
3	W. Sumatra	10	W. Java	17	East Timor	24	S. Sulawesi
4	Riau	11	C. Java	18	W. Kalimantan	25	SE. Sulawesi
5	Jambi	12	D. I. Yogyakarta	19	C. Kalimantan	26	Maluku
6	S. Sumatra	13	E. Java	20	S. Kalimantan	27	Irian Jaya
7	Bengkulu	14	Bali	21	E. Kalimantan		

Figure 2.1 AIDS case rate by province: cumulative cases per 100,000 people as at 31 August 1999 (source: UNAIDS (Jakarta) (2001)).

Table 2.2 West Papua: range in prevalence rates of HIV in high- and low-risk populations at the end of 2000

Type of population	HIV prevalence range found in surveillance sites
Female sex workers	0–26.5
Clients	0–2.75
Fishermen	0–2.8
Transport drivers	0–2.6
Factory workers	0–1.2
General population	0–1.16
Students	0–0.6
Pregnant women	0–0.25

Source: Directorate General of Communicable Disease Control and Environmental Health (2001).

highest incidence was among female sex workers. There was a considerable regional variation in the concentration of female sex workers, with the highest concentration in Merauke (Figure 2.2). However, substantial infection was present in all the sites, especially Sorong, Biak and Manokwari. The bulk of HIV positive cases have involved heterosexual transmission (94.9 per cent), with small proportions through homosexual (2.2 per cent) and perinatal (2.9 per cent) transmission. One aspect of the infection in West Papua is that there is a larger proportion of AIDS cases that are female (35.5 per cent) than nationally. Some 39 per cent are aged twenty to twenty-nine and another 31 per cent are aged thirty to thirty-nine.

Aside from West Papua, in Indonesia the largest number with HIV/AIDS are in Jakarta which has more than one-third of reported cases. It also has the second highest notification rate in the country, slightly less than one-third of that in West Papua. Indonesia's dominant urban centre has concentrations of many of the high-risk groups for HIV infection, including large numbers of female sex workers, a substantial gay population, large numbers of single and unaccompanied migrant workers, and intravenous drug users (IDUs).

The province of Riau also has an above-average prevalence of HIV, and population mobility has played a significant role in this. In the Riau archipelago, there has long been a great deal of movement between the various islands – Sumatra, Singapore and Peninsula Malaysia. There is also a well-established commercial sex industry and pattern of Indonesian female sex workers serving males from Singapore. This has been partly through the women going to Singapore (usually undocumented) to conduct business overnight and then return. In more recent times the entertainment industry in several of the islands near Singapore, especially Batam, has become highly organized to serve day-tripping Singaporeans (and increasingly Malaysians), and the sex industry is a major part of this. Batam is the main centre but the industry is also thriving on several other islands (Jones *et al.* 1998).

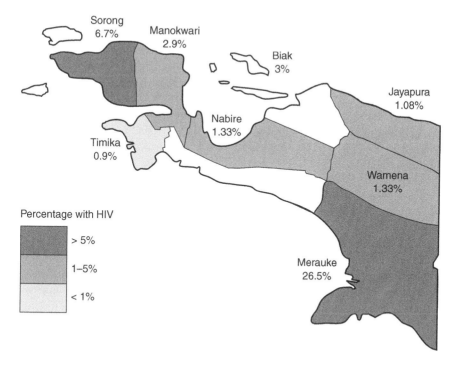

Figure 2.2 West Papua: prevalence of HIV among female sex workers in Sentinel Surveillance Sites, 1997 to 2000 (source: Directorate General of Communicable Disease Control and Environmental Health (2001)).

Among the remaining provinces, there is a significantly high reporting of HIV/AIDS cases and rates per 100,000 population. Bali, for example, has the third highest notification rate (0.8/100,000) and the sixth largest number of reported cases among the provinces. Irwanto (2001) reported that in December 2000 HIV testing was conducted by the Bali Health Department on 187 prisoners in Kerobokan Prison in Denpasar. Of these, 160 were drug users and sixty-six IDUs and of these, thirty-four were HIV positive. The reasons for the above-average incidence of HIV infection in Bali are unclear. Setiawan (2002: 24) found high levels of use of commercial sex workers (CSWs) by men in rural eastern Bali. Some have suggested that it could be associated with high-risk activities linked to the high level of tourist activity in the province, especially in the substantial local drug and sex industries. However, it is also true that there are perhaps higher levels of surveillance in Bali, and so it is more a matter of a greater proportion of infected people being detected.

Among the other Indonesian provinces, North Sulawesi has the fifth highest rate of reported cases per 100,000. The reasons for this elevated level are unclear, but the province is an important source of female sex workers from elsewhere, including East Kalimantan, East Malaysia and the Philippines. In South Sumatra, Palembang city is the third largest outside of Java. It has a thriving sex industry and is a major target of migration (Jones *et al.* 1998). In East Java, Surabaya is a major centre of brothels in Indonesia (Jones *et al.* 1998; Steele 1981), as well as a major source of CSWs throughout Indonesia. Surjadi (2001) reports that CSWs in Surabaya in 1996, 1998 and 2000 reported sexually transmitted infection rates of 37.0, 60.5 and 31.3 per cent. Finally, West Java is the largest province in Indonesia, and although the sex industry in the main city (Bandung) is not as substantial as in Surabaya, some areas of West Java (especially *kabupaten* Indramayu and, to a lesser extent, Cirebon, Subang and Purwakarta) are origin areas for CSWs.

The age distribution of people reporting infection is depicted in Figure 2.3. Almost half (46.2 per cent) were aged between twenty and twenty-nine, while 25.9 per cent were aged between thirty and thrity-nine, 8.6 per cent aged forty to forty-nine, and 8.1 per cent aged fifteen to nineteen years. Among the cases of HIV infection, 38.2 per cent were female and among those with AIDS, 19.3 per cent were female. Some 18.4 per cent were foreigners. The modes of transmission among the reported cases were heterosexual (70.5 per cent), homosexual/bisexual (9.6 per cent), and injecting drug user (18.3 per cent).

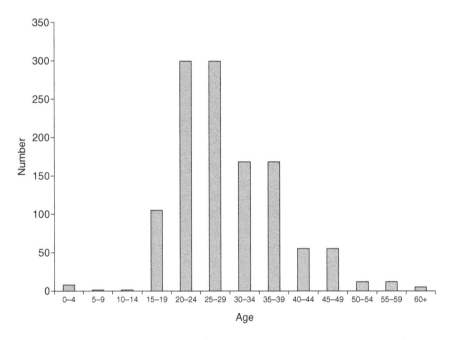

Figure 2.3 Indonesia: age structure of the population reported with HIV infection, November 2000 (source: Directorate General of Communicable Disease Control and Environmental Health (2001)).

Internal population mobility in Indonesia

Indonesians are stereotyped as having low levels of population mobility and this, to some extent, is confirmed by census data on migration. In the most recent population census, for example, only one in ten Indonesians was classified as a migrant. Although Table 2.3 shows that this proportion doubled in the past three decades, it still suggests a low level of mobility. However, the census data mask the bulk of Indonesian movement. There are estimated, for example, to be five migrations within provinces for every one between them, and it is only the latter that is detected in the census (Hugo 1982). More importantly, the census captures only more or less permanent migrations and ignores moves that do not involve a permanent change in the usual place of residence. Non-permanent circulation influences many more Indonesians than permanent displacement (Hugo 1975, 2002).

The massive increase in personal mobility of Indonesians in the past five decades is demonstrated in Table 2.4, which shows that over this period the number of Indonesians per registered motor vehicle declined from 1,507 to eleven. This has greatly expanded the area over which many Indonesians travel, especially to seek work within Indonesia. Non-permanent internal circular migration is substantially greater in scale than permanent migrations. Circular migration usually has several characteristics which are important from the perspective of examining the potential spread of HIV/AIDS. Migrant workers are drawn selectively from young adults, especially males, and migration involves separation from spouses and/or parents for extended periods of up to two years. Migrants are particularly directed to places where the commercial sex industry is well established, including urban areas, plantations, remote construction sites, oil rigs, mining sites,

Table 2.3 Indonesia: measures of migration 1971–2000

Year	Migration measure	Male	Female
1971	Per cent ever lived in another province	6.29	5.06
1985	Per cent ever lived in another province	8.37	7.29
1985	Per cent intra-province migrants	7.04	6.75
1985	Per cent five-year migrants:		
	inter-provincial	2.07	1.85
	intra-provincial (between *kabupaten*)	1.97	1.89
1990	Per cent ever lived in another province	10.62	9.03
1990	Per cent five-year migrants: inter-provincial	3.54	3.12
1995	Per cent ever lived in another province	11.19	10.00
1995	Per cent five-year migrants: inter-provincial	2.46	2.37
2000	Per cent ever lived in another province	10.56	9.57
2000	Per cent five-year migrants: inter-provincial	3.16	2.96

Source: Indonesian Censuses of 1971, 1990 and 2000; Intercensal Surveys of 1985, 1995 and 2000.

Table 2.4 Indonesia: number of persons per motor vehicle 1950–2000

Year	Persons/motor vehicle (including motor cycles)	Persons/motor vehicle (excluding motor cycles)
1950	1,507	1,691
1956	446	735
1961	263	447
1966	191	375
1971	129	300
1976	63	193
1980	38	123
1985	24	80
1990	20	64
1995	15	47
2000	11	39

Source: Biro Pusat Statistik publications on motor vehicles and length of roads.

border industrial areas and forestry sites. Migrants often earn relatively large amounts of cash, are often far removed from traditional constraints on behaviour, and almost all commercial sex workers are internal circular migrants. This type of mobility involves several millions of workers. The crucial point here is that circular mobility rarely involves entire family units, unlike permanent migration which tends to involve family groups.

In recent times there have been some other developments in internal population mobility which may also have implications for the spread of HIV/AIDS. The first is the proliferation of large factories in the Jabotabek area, and to a lesser extent in areas such as Batam and Bandung in West Java. This has involved the recruitment of many young workers, especially women, who live in dormitories, barracks and crowded boarding-houses. Further, the 1997 financial crisis displaced large numbers of these urban workers, some of whom returned to their rural places of origin, but many others of whom moved into the informal sector and swelled the numbers working in the informal commercial sex industry (Hugo 2001). Another development has been the displacement of more than 1.3 million Indonesians by the outbreak of violence in several outer island locations after 1998. Many of these internally displaced persons have been accommodated in 'temporary' camps. These places have a history in Africa of fostering the extent and spread of HIV infection.

Despite the above, there is no evidence to suggest that circular migrants in Indonesia have a higher rate of HIV infection than do non-migrants. This is because migrant workers have not been included among the sites for sentinel surveillance activities and the collection of data on HIV infection. It must be re-emphasized that it is inappropriate to assume that migrants *per se* have a greater vulnerability to HIV infection. The evidence is that many are placed in contexts where they are at higher risk of infection than would be the case if they had remained in their home areas. Circular mobility is occurring on a massive scale in Indonesia, and in Java few rural households are not influenced by it. This is a concern not only because of the high level of individual labour mobility in Indonesia, but also because of the nature of that mobility and the extent to which it puts movers in destination situations that are potentially high-risk areas.

International migration

Indonesia has a more limited diaspora of citizens who have moved more or less permanently to other, especially developed, countries than is the case for many of its Asian neighbours. However, it is one of the world's largest suppliers of international labour migrants (Hugo 2003). This movement shares many of the characteristics of the internal circular migration examined in the last section. Table 2.5 presents an estimate of the current stock of Indonesian contract workers in foreign destinations. Estimates are difficult because up to half of the movement is undocumented and does not enter into official data. Nevertheless, it may be asserted that more than two million Indonesian workers at any one time are working in other Asian or Middle Eastern countries. Moreover, the numbers appear to be increasing since the crisis years of the late twentieth century.

As is the case with internal migration, there are few or no data regarding the extent of HIV infection among these migrant workers. However, there are a number of elements to suggest that many such overseas contract workers (OCWs) are in vulnerable contexts in their destinations. Most OCWs, the majority of whom come from rural areas and have limited skills, move without their families and may be separated from family and friends at their destination usually for a period of around two years. Women are a major

Table 2.5 Indonesia: estimated stocks of overseas contract workers around 2000

Destination	Estimated stocks	Source
Saudi Arabia	425,000	Indonesian Embassy Riyadh
United Arab Emirates	35,000	*Asian Migration News*, 30 April 1999
Malaysia	1,376,000	*Migration News*, October 2001
Hong Kong	66,000	*Migration News*, December 2002
Singapore	70,000	*Asian Migration News*, 5 May 1999
Taiwan	90,739	*Migration News*, April 2003
South Korea	25,473	*Asian Migration News*, December 2002
Japan	3,245	*Asian Migrant Yearbook*, 1999: 128
Philippines	26,000	*SCMP*, 10 December 1998
Brunei	2,426	*Asian Migration Yearbook*, 1999: 125
Other	20,000	DEPNAKER
Total	2,139,883	

element in the migration and, in fact, dominate in the official migration. They are employed predominantly as household domestic workers which can, and does, expose many to the threat of exploitation. There is a significant movement of Indonesian women to work in the commercial sex industry in Malaysia (Jones 1996), and there is increasing evidence of trafficking of women from Indonesia to elsewhere in Southeast Asia (Jones 2000). Ultimately, there is little official support for OCWs once they are in a foreign country. Perhaps more than a half of OCWs are not legally resident in their destination country which exposes them to the risk of exploitation. There is a thriving commercial sex industry in many of the destinations of OCWs, exacerbated by the fact that OCWs often have disposable cash income.

As indicated above, there are few data to suggest that Indonesian OCWs have a higher rate of HIV infection than do non-migrants at the destination or origin. While it is illegal to insist that workers be compulsorily tested for HIV before or after migration (*Asian Migration News*, December 2002: 1–15), many destination countries and employers do insist on pre-departure testing and periodic testing while they are in the destination (e.g. Singapore). No data are available regarding either of these activities. A Red Cross study in Nepal found that 10 per cent of OCWs going to India became HIV positive due to unprotected sex (*Asian Migration News*, April 2003: 16–30). In the Philippines it has been found that 30 per cent of all HIV positive notifications are former OCWs (*Asian Migration News*, June 2003: 1–15) but no similar information is available for Indonesia. One set of data released relates to Brunei which compulsorily tests incoming Indonesian OCWs; of 33,865 tested, only one was HIV positive (Parida 2000). It should be noted that Indonesia has an extremely high rate of premature return among its official overseas workers (Hugo forthcoming), but it is not known what proportion of these returnees has been found to be HIV positive.

While the bulk of Indonesian CSW international labour migration is to Malaysia, it does occur to other areas. In 2003 (*Asian Migration News*, May 2003: 16–31), it was reported that 100 Indonesian migrant workers working as prostitutes in several cities in Saudi Arabia had been arrested and were being deported. Another report (*Asian Migration News*, June 2003: 16–30) was that 1,500 young Indonesian women were trapped in the sex business in Abu Dhabi after being tricked into moving there with the promise of well-paid jobs as factory workers or domestic workers.

The nexus between the commercial sex industry, mobility and HIV/AIDS

The previous two sections have examined the strong relationship in Indonesia between the commercial sex industry and population mobility. Indonesia has an estimated 190,000 to 270,000 female sex workers and there are seven to ten million men who are regular clients. Half of these men have stable partners or are married (Ministry of Health, Republic of Indonesia 2002: 10).[2]

There are several elements in this relationship. First, almost all commercial sex workers are themselves circular migrants. They do not practise their business in their home communities. Second, mobile groups are among the heaviest users of the services of the commercial sex industry. Results of behavioural surveillance surveys in several Indonesian towns show that more than 50 per cent of highly mobile males bought sexual services in the past year (Ministry of Health, Republic of Indonesia 2002: 11). Finally, the Indonesian commercial sex industry is concentrated in areas where there are large numbers of circular migrants – in cities, border-crossing points, remote clusters of employment in construction, mining, plantations, at tourist destinations, along transport routes, and in transport centres and ports.

Commercial sex workers in Indonesia are frequently placed in powerless situations where even if they wish to use condoms, the prevailing anti-condom attitudes among male customers prevent them from using them. The data show a relatively low use of condoms among CSWs in Indonesia with less than 10 per cent systematically wearing condoms to avoid being infected (Ministry of Health, Republic of Indonesia 2002: 10). However, these studies have been undertaken in the *lokalisasi* or government-condoned red light districts. The fact is that much prostitution in Indonesia is practised outside of these zones and is totally unregulated, and almost certainly the use of condoms in these areas is even lower.

The potential for the spread of HIV infection through the strong interaction between migrant workers and the commercial sex sector is demonstrated in Figure 2.4. The infection can readily be passed from

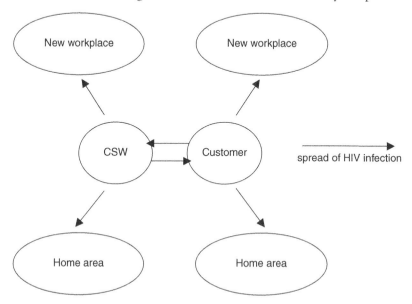

Figure 2.4 Model of potential spread of HIV/AIDS in Indonesia through the commercial sex industry.

customer to CSW, and vice versa. However, as Figure 2.4 shows, both CSWs and customers then have a substantial capacity to spread the infection. First, both can take it back to their home area and pass it on, presumably to their partners,[3] as shown in Figure 2.5. However, a strong feature of the mobility of internal migrant workers and CSWs in Indonesia is that they do not always tend to go back to the same destination to work. The existence of established circuits of circular migration among CSWs occurs not only within Indonesia but also in international sex worker migrations (Brockett 1996). Hence there are multiple diffusion effects as both CSW and customer become agents of diffusion.

The important point here is that infection can and does spread quite rapidly due to the patterns of mobility. A typical case of CSW mobility was found in a study in West Papua. The first panel in Figure 2.6 depicts the locations at which CSWs were interviewed – areas associated with mining, transport and urban activity. The second panel depicts their home places, and it will be seen that the majority are from places well known as sources of CSWs, such as East Java and North Sulawesi. The third panel depicts the other parts of Indonesia where the sampled CSWs have worked, and it will be noted that they have moved throughout the archipelago.

The situation in Riau illustrates the complex relationship between mobility, HIV/AIDS and the commercial sex industry. It was shown earlier that Riau has one of the highest levels of incidence of HIV/AIDS. Batam had 112

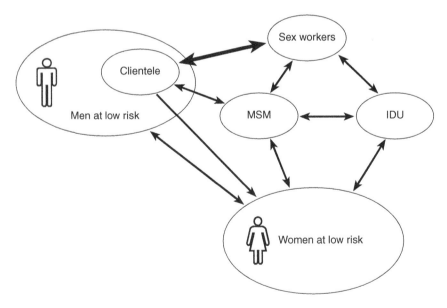

Figure 2.5 Potential increase in means of HIV transmission in Indonesia, from one risk group to another, through sexual activities without condoms between the groups in question (source: Ministry of Health, Republic of Indonesia (2002)).

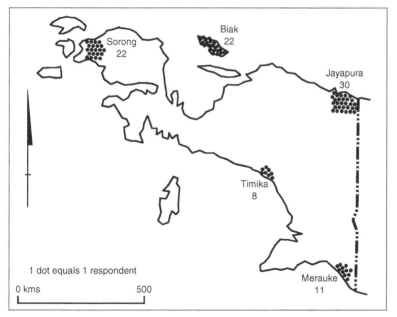

1 Location of commercial sex workers interviewed

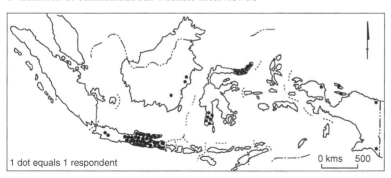

2 Commercial sex workers' home areas

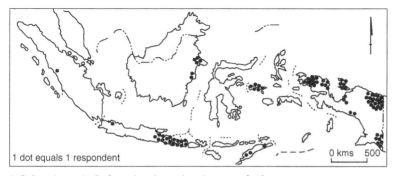

3 Other Areas in Indonesia where they have worked

Figure 2.6 Migration of commercial sex workers to and from West Papua (source: Wiebel and Safika (n.d.)).

cases reported in 2002 although this is seen as not indicative of the true extent of the disease (*Asian Migration News*, January 2003: 16–31). With a population of 520,000, Batam is one of Indonesia's fastest growing regions. Influenced by its location close to Singapore, there has been substantial tourism development. It is visited by around 3,000 foreign tourists (mostly Singaporeans) each month (*Asian Migration News*, June 2003: 18–30), many of whom are attracted by its sex industry. It is not only Indonesian sex workers, however, who have been attracted by the booming sex sector on Batam. Around 150 sex workers from Thailand, Taiwan, China, Hong Kong and several European countries have been reported to be active in parts of Batam (*Asian Migration News*, December 2002: 1–15).

In Riau the proliferation of the sex industry has not only been a function of serving Singaporean and Malaysian tourists. For twenty years Batam has been a major Indonesian government development zone; its close proximity to Singapore has been capitalized upon by attracting a great deal of manufacturing activity through offering a range of incentives. It is a free trade zone and it has also served as an 'overspill' area for Singapore, from which many activities have been squeezed due to limited space, high costs or regulations. The pig-raising industry, for example, has been forced out of Singapore by health regulations and is now based mainly in Indonesia. This has seen a proliferation of factories in Batam which have not been able to access sufficient labour from the local population. As a result, there has been large-scale movement to Batam from elsewhere in Indonesia, mainly Java and Sumatra. Among the migrants, singles or unaccompanied married persons have predominated (both males and females) and the already substantial sex industry has expanded to meet the new demand from this group.

Policy implications

With the exception of a few restricted geographical areas, HIV/AIDS has not reached epidemic proportions in Indonesia. Nevertheless, there is substantial potential for the high levels and distinctive characteristics of population mobility in Indonesia to facilitate the rapid spread of infection within a short period of time. There are a number of implications relating to both preventive action against, and treatment of, HIV/AIDS. Above all there needs to be recognition of the existing and potential significance of population mobility in relation to HIV/AIDS. This recognition has come in Africa, but in many cases too late to stop the spread of the disease (Hsu and Du Guerny 2000). In Thailand full appreciation of the nexus between mobility, the sex industry and HIV infection has been instrumental in the successful introduction of preventive interventions which have arrested growth in the rates of infection. In Indonesia this recognition is coming slowly.

An important point relates to the fact that highly mobile groups in Indonesia tend to be spatially concentrated in particular areas. There is then an urgent need to clearly identify which highly mobile groups are at risk, and in what areas they are concentrated. Such groups overwhelmingly comprise migrant workers of various kinds and they may be identified by the types of work in which they engage.[4] There are four key groups.

First, there are those in itinerant occupations. There is substantial literature which demonstrates the high levels of involvement of groups such as truck drivers, fisher people, sea farmers, itinerant traders, bus drivers and so on with the commercial sex sector, and hence, high vulnerability to HIV/AIDS.[5]

Second, there are workers involved in circular migration. There is less knowledge of the extent to which these workers, who leave their usually rural-based families for extended periods to work in urban areas or other sites where there are abundant job opportunities, such as often remote construction sites, plantations, mines and forestry operations, have a higher incidence of HIV infection. However, what is known is that they make use of the commercial sex industry more than men who remain with their families, and the chances are strong that they are being exposed to a significant risk of HIV infection.

Commercial sex workers comprise the third category of highly mobile groups. There is growing recognition in Asia of the substantial size, and widespread use, of the commercial sex sector (Lim 1998). There is less recognition, however, that they are also generally circular migrants who practise their occupation a considerable distance from their home place and families. They are obviously at high risk due to the nature of their work, but this is exacerbated in Indonesia by the fact that they lack power, so that while they may wish to use condoms, their customers are able to insist that they do not. Moreover, in Indonesia there is an entrenched anti-condom attitude among men.

The final category of highly mobile people comprises the army and police forces. This is a large group in Indonesia which is posted away from their families for extensive periods, as was the case during recent widespread conflict and unrest in several peripheral parts of the archipelago (Hugo 2002). There are also a number of other elements which may be relevant, such as an ethos of risk-taking and unequal power relations with the communities into which they are posted.

Many millions of Indonesians fall into these four occupational categories, and while we must be careful not to stereotype these groups as having higher than average levels of HIV infection *per se,* they are often placed in situations where they are at risk of infection. From a policy and programme perspective, the fact that these groups are not fully dispersed throughout Indonesia, but are concentrated in particular ecological niches, is relevant. Such spatial concentration has a number of potential advantages to policymakers and planners.

First of all, those spatial concentrations mean that sentinel surveillance activities for measuring the level of HIV infection and the impact of intervention activities can be located in those areas rather than throughout the entire population. In addition, there may be a focus on such areas in information and preventive activities as well as in locating facilities and resources for the treatment of those infected with HIV. Clearly, in a situation of scarce resources, the ability to concentrate on a few areas means that there can potentially be maximum impact for the expenditure available. This has been shown in Thailand where the concentrations of activities such as the 100 Percent Use of Condom Programmes in concentrations of groups like truck drivers and commercial sex workers has been instrumental in curbing the exponential expansion of infection. In Indonesia, the main types of concentration are: ports and other important transhipment points and transport hubs as well as along major transport routes; major urban centres, especially fast-growing cities; major constructions sites, especially those in more remote areas; border areas; other areas where large numbers of contract workers are employed (e.g. plantations, forestry enterprises, mines, oil extraction areas, remote factories and processing plants).

In such areas it is important to not only concentrate on the government defined and sanctioned areas of prostitution or *lokalisasi* (Jones *et al.* 1998), but also the informal commercial sex industry areas, since the bulk of migrant workers tend to patronize the lower cost end of the industry which is generally outside the *lokalisasi*.

In the initiation and targeting of prevention and intervention strategies it is necessary to recognize that population mobility involves not only a single destination place but also a system involving origins, multiple destinations, transit points and transport routes. The migrant workers, and the people with whom they interact, may be found in all of these locations, so intervention efforts should not focus purely on a single part (usually the destination) of these systems. Preventive activity and information programmes, for example, need to be initiated, coordinated and integrated across these mobility systems. It needs to be stressed that the intervention efforts should not be directed at the migrant worker groups *per se*, but at risk reduction in the areas where these groups are concentrated. Intervention needs to focus on the risk behaviour. Intervention efforts also need to be initiated with full recognition of the human rights of all involved, including the migrant workers. It may also be the case that the mobility of the groups involved will require modification of the way in which interventions are made, and the types of support materials and services provided.

Vulnerability to HIV infection among international labour migrants has some similar policy implications to those discussed for internal circular migrants, but there are also some additional considerations. The organization of coordinated origin/destination programmes is much more difficult when more than one country is involved, but international coordination is necessary. It may be that NGOs are more able to effect such cooperation than are governments. A starting point to such cooperation may be at border-crossing points where there tend to be urban areas on both sides of the border. Although the concentration of risk-taking behaviour may be in one or other of the places, it will involve people from both sides of the border. Another complicating issue in international labour migration is the fact that many are undocumented, especially among those engaged in the most 'at-risk' occupations. This makes it even less likely that they are able to access relevant information and prevention support both in origin and destination areas. It is crucial that these activities are inclusive of both documented and undocumented overseas contract workers.

It has been argued elsewhere (Hugo forthcoming) that one of the major shortcomings of Indonesia's international labour migration programme is the limited quality, quantity and relevance of training and pre-departure preparation provided to OCWs. While this is a specific requirement of official international labour migration in practice, most OCWs are only partially prepared for their overseas working experience, and this is undoubtedly a major factor in the very high rates of premature return before contracts are fulfilled among Indonesian OCWs. Undocumented workers do not receive any formal pre-departure training and preparation, although many are prepared informally, since they often move with relatives and friends who had previously moved. Certainly in Indonesia, urgent priority should be given to the development and introduction of comprehensive, targeted and appropriate health modules to be included in the training given to all OCWs about to leave Indonesia. A model for such a module was introduced into Sri Lanka in 2001 (*Asian Migration News*, June 2001: 1–15). The programme covers modes of transition and prevention of sexually transmitted diseases and HIV/AIDS, personal hygiene and medical fitness, family health and health problems in the workplace. In the case of women who make up the majority of official migrants, there is also a complementary need for training in a range of defensive strategies to combat exploitation of various kinds. While these need to be introduced into training programmes for formal OCWs, there should be other activities aimed at undocumented OCWs. This would be best done through NGOs and require the development of informal delivery mechanisms.

The introduction of effective prevention and information programmes for commercial sex workers is of crucial importance not only within *lokalisasi* but also among the CSWs who operate outside them. Similar activity among their clients, both migrants and non-migrants, is also crucial to limiting the spread of the disease. Programmes which empower the CSWs to not only want to practise safe sex, but to be able to insist on it from their clients, are especially important in the Indonesian context.

There is a real danger of unjustly stereotyping and stigmatizing migrant workers as prone to HIV infection and as agents of the spread of the disease. This can lead to discrimination, denial of human rights and mistreatment. Indeed, there is already evidence of this among Indonesian OCWs in Malaysia. A less obvious factor has been raised by the Directorate General of Communicable Disease Control and Environmental Health (2001: 7) who has stated: 'This kind of mobility ... gives local communities an excuse to ignore the importance of risk behaviours that are taking place in their midst, by claiming commercial sex is an "outsiders" problem. This gives a false sense of security.'

This is a real issue in Indonesia. To many, HIV/AIDS is a disease which only affects people on the margins of mainstream society, and hence is not of concern to health providers, or to the public more generally. Thus HIV/AIDS was initially, and in some cases still is, branded in Indonesia as associated with foreign tourists and fisher people who visit Indonesia. Similarly, if it is ascribed to CSWs, drug users and other marginal groups there is a real danger of complacency, such that the disease is not given its proper priority in health provision and public awareness.

Conclusion

Recognition of the potential of HIV/AIDS to attain epidemic proportions in Indonesia has been slow in coming. Even now there are many myths regarding vulnerability to the disease. The high levels, and distinct nature of Indonesian population mobility, especially that involving workers, is one of the most important elements which can potentially lead to a rapid spread of the disease within Indonesia. The present need, however, is not only for the development and initiation of policies and programmes which are directed towards prevention of infection among migrant workers, although these are very pressing. More pressing is the need for research. There is little or no detailed knowledge in Indonesia of either the HIV/AIDS risk behaviour in which Indonesian migrant workers engage, or the extent of infection among various types of migrant workers. This research task is urgent if Indonesian labour migrants are to be empowered against infection by being provided with appropriate, timely and effective information, and preventive policies and programmes.

There are, however, a number of interventions which can be recommended immediately. 'Hot Spots' within Indonesia where there are concentrations of CSWs and migrant workers need to be clearly identified and targeted with information and prevention programmes. This can be done quickly and cheaply. Further, some migrant worker groups need to be included in the sentinel surveillance programme. It should be mandatory for employers of large numbers of migrant workers to provide a full range of HIV/AIDS information and services. Modules on HIV/AIDS awareness raising, information and prevention need to be mandatory in all compulsory training programmes for OCWs. There also need to be innovative programmes of HIV/AIDS awareness raising, information and prevention targeted to origin areas of undocumented OCWs – provided largely through NGOs.

Beyond this, cooperative activities regarding awareness raising, information and prevention of HIV/AIDS and procedures and assistance for OCWs infected with HIV/AIDS need to be negotiated between the Indonesian government and all major destination countries of OCWs. All overseas employment companies in Indonesia should provide a full range of HIV services to the OCWs they recruit and place. NGOs concerned with HIV/AIDS need to be linked with NGOs who focus on the concerns of OCWs. The waiting times for deployment and at border-crossing points need to be reduced to avoid the concentrations of large numbers of idle migrant workers in close proximity to sex industry concentrations.

Currently, Indonesia has a low level of HIV infection but there are several indications that there is no room for complacency. There are areas where infection levels are high and indications that the rate of infection is increasing. Moreover, there are several conditions in Indonesia, which in parts of Africa were conducive to the rapid spread of the disease. Among these, the very high levels of mobility, particularly that of young people moving without their families, is especially important. It is imperative that the knowledge of the nexus between high mobility, the commercial sex sector and HIV infection be used to inform the design and targeting of information, prevention and sentinel surveillance programmes. On the other hand, there is a need to guard against the stereotyping of movers and commercial sex workers as being responsible for the spread of HIV. Such stigmatization is not only incorrect but is doubly damaging. First, it suggests wrongly that purely by moving, people are at greater risk of HIV infection, when in fact it is the behaviour of certain types of movers which actually puts them at risk and not the mobility *per se*. Second, it reinforces the unfortunate practice in Indonesia of dismissing HIV as an infection confined to peripheral groups in society and not of significance in mainstream society. Such attitudes are a barrier to the mobilization of resources and commitment which is needed to prevent the already significant tragedy of HIV from becoming more widespread in Indonesia than it is already.

Notes

1 The only other means of transmission is perinatal transmission from mother to child.
2 On a smaller scale commercial sex is also being provided by male sex workers and transvestites (Ministry of Health, Republic of Indonesia 2002: 10).
3 It has been reported in a few areas of Jakarta that HIV transmission has affected the parties of those in at-risk groups and 3 per cent of 500 pregnant women who volunteered for testing have contracted HIV (Ministry of Health, Republic of Indonesia 2002: 17).
4 These are considered in more detail in Hugo (2001).
5 It is clear too that these groups are not only at high risk of HIV infection through the sex sector but also because there is high use of drugs and needle sharing (Klanarong 2003).

References

Appleyard, R. and Wilson, A. (eds) (1998) 'Special Issue: Migration and HIV/AIDS', *International Migration*, 36, 4.

Asian Migrant Center (1999) *Asian Migrant Yearbook 1999,* Asian Migrant Center Ltd, Hong Kong.

Bickers, C. and Crispin, S.W. (2000) 'Asia sets its sights on an AIDS breakthrough', *Far Eastern Economic Review*, 7 December: 38–43.

Biro Pusat Statistik. *Statistik Kendaraan Bermotor dan Panjang Jalan* (Vehicles and Length of Road Statistics), Biro Pusat Statistik, Jakarta, various issues.

Brockerhoff, M. and Biddlecom, A.E. (1999) 'Migration, sexual behaviour and the risk of HIV in Kenya', *International Migration Review*, 33, 4: 833–56.

Brockett, L. (1996) 'Thai sex workers in Sydney', unpublished MA thesis, Department of Geography, University of Sydney.

Brown, T. (2002) 'The HIV/AIDS epidemic in Asia', *Asia-Pacific Population and Policy*, No. 60, East–West Center, Honolulu.

Caldwell, J.C., Anarafi, J.N. and Caldwell, P. (1997) 'Mobility, migration, sex, STDs and AIDS', in G. Herdt (ed.) *Sexual Cultures and Migration in the Era of AIDS*, Oxford: Clarendon Press, pp. 41–54.

Decosas, J., Kane, F., Anarfi, J.K., Sodji, K.D.R. and Wagner, H. (1995) 'Migration and Aids', *Lancet*, 346: 826–29.

Directorate General of Communicable Disease Control and Environmental Health (2001) *HIV/AIDS and Other Sexually Transmitted Infections in Indonesia: Opportunities for Action*, National Report Draft 3, Ministry of Health, Republic of Indonesia.

Gardner, R. and Blackburn, M.S. (1996) 'People who move: new reproductive health focus', *Population Reports*, Series J, 45: 1–7.

Haour-Knipe, M. and Rector, R. (1996) *Crossing Borders: Migration, Ethnicity and AIDS*, London: Taylor & Francis.

Herdt, G. (ed.) (1997) Sexual Cultures and Migration in the Era of AIDS: Anthropological and Demographic Perspectives, Oxford: Clarendon Press.

Hsu, L. and Du Guerny, J. (2000) 'Population movement, development and HIV/AIDS: looking towards the future', in UNDP South East Asia HIV and Development Project, *Population Mobility in Asia: Implications for HIV/AIDS Action Programmes*, UNDP, South East Asia HIV and Development Project, Bangkok, pp. 1–7.

Hugo, G.J. (1975) 'Population mobility in West Java, Indonesia', unpublished Ph.D. thesis, Department of Demography, The Australian National University, Canberra.

Hugo, G.J. (1978) *Population Mobility in West Java*, Yogyakarta: Gadjah Mada University Press.

Hugo, G.J. (1982) 'Sources of internal migration data in Indonesia: their potential and limitations', *Majalah Demografi Indonesia* (Indonesian Journal of Demography), 17, 9: 23–52.

Hugo, G.J. (2001) *Indonesia: Internal and International Population Mobility: Implications for the Spread of HIV/AIDS*, UNDP South East Asia HIV and Development Office, UNAIDS and ILO, Indonesia.

Hugo, G.J. (2002) 'Pengungsi – Indonesia's internally displaced persons', *Asian and Pacific Migration Journal*, 11, 3: 297–331.

Hugo, G.J. (2003) 'Labour migration, growth and development in Asia: past trends and future directions', paper prepared for ILO Regional Tripartite Meeting on

Challenges to Labour Migration Policy and Management in Asia, Bangkok, 30 June to 2 July.

Hugo, G.J. (forthcoming) 'Information, exploitation and empowerment: the case of Indonesian overseas contract workers', *Asian and Pacific Migration Journal*.

Indonesian Directorate of Direct Transmitted Disease Control (2001) 'Penyelamatan Irian Jaya Terhadap Bahaya HIV/AIDS', proposal prepared by Indonesian Directorate of Direct Transmitted Disease Control.

Irwanto (2001) *Rapid Assessment and Response – Indonesia: Jakarta, Bandung, Yogyakarta, Surabaya, Denpasar, Medan, Ujung Pandang, Marado*. In cooperation with AusAID, USAID, WHO, UNAIDS, Ford Foundation, PATH and KPA Nasional, Presentation, National AIDS Commission, Denpasar, Bali, 2–4 May.

Jones, G.W., Sulistyaningsih, E. and Hull, T.H. (1998) 'Prostitution in Indonesia', in L.L. Lim (ed.) *The Sex Sector: The Economic and Social Basis of Prostitution in Southeast Asia*, Geneva: International Labour Office, pp. 29–66.

Jones, S. (1996) 'Women feed Malaysian boom', *Inside Indonesia*, July to September, 47: 16–18.

Jones, S. (2000) *Making Money Off Migrants: The Indonesian Exodus to Malaysia*, ASIA 2000 Limited, Hongkong and CAPSTRANS, University of Wollongong.

Kaldor, J. (2000) 'Biomedical responses to the AIDS epidemic', paper presented at National Academies Forum, Every Eight Seconds: AIDS Revisited, Canberra, November.

Klanarong, N. (2003) 'Female international labour migration from Southern Thailand', unpublished Ph.D. thesis, Population and Human Resources, Department of Geographical and Environmental Studies, The University of Adelaide.

Lim, L.L. (ed.) (1998) *The sex sector: the economic and social bases of prostitution in Southeast Asia*, Geneva: International Labour Office.

Ministry of Health, Republic of Indonesia (2002) 'Increasingly evidence threat of HIV/AIDS in Indonesia, calls for more concrete prevention efforts', Ministry of Health, Republic of Indonesia, Jakarta, Draft.

Parida, S.K. (2000) 'Sexually-transmitted infections and risk exposure among HIV positive migrant workers in Brunei Darussalam', in UNDP South East Asia HIV and Development Project, *Population Mobility in Asia: Implications for HIV/AIDS Action Programmes*, UNDP, South East Asia HIV and Development Project, Bangkok, pp. 26–32.

Prothero, R.M. (1965) *Migrants and Malaria*, London: Longman.

Prothero, R.M. (1977) 'Disease and mobility: a neglected factor in epidemiology', *International Journal of Epidemiology*, 6, 3: 259–67.

Setiawan, I.M. (2002) 'Fighting Aids in Indonesia', *Far Eastern Economic Review*, 28 March: 24.

Skeldon, R. (2000) *Population Mobility and HIV Vulnerability in South East Asia: An Assessment and Analysis*, UNDP, South East Asia HIV and Development Project, Bangkok.

Smith, C.J. and Yang, X. (2002) 'Is the "Floating Population" to blame?: examining the connection between temporary migration and the HIV/AIDS epidemic in China', paper presented at the International Conference on Population Geographies, St. Andrews, Scotland, 19–23 July.

Steele, R. (1981) *Origin and Occupational Mobility of Lifetime Migrants to Surabaya, East Java*. Unpublished Ph.D. thesis in geography, Australian National University, Canberra.

Surjadi, C. (2001) Prevalensi PMS Pada PSK di Jakarta, Surabaya dan Manado Serta Peran Pria Sebagai Provider dan Pasien. Dibakan pada Seminar Sehari Peningkatan Pesan dan Tanggung Jawab Laki-laki Dalam Upaya Menghambat Epidemi HIV/AIDS di Indonesia, Jakarta, 25 January.

UNAIDS (Jakarta) (2001) *HIV/AIDS Report 2000 Indonesia*, Jakarta, Mimeo.

Walzholz, G. (2003) 'Migration and HIV/AIDS in Asia', paper presented at ILO Regional Tripartite Meeting on Challenges to Labour Migration Policy and Management in Asia, Bangkok, Thailand, 30 June to 2 July.

White, R., Kintu, P.B., Orroth, K.K., Korenromp, E.L., Mwita, W., Gray, R.H., Kamali, A., Sewankambo, N.K., Wawer, M.J., Whitworth, J.A.G., Grosskurth, H., Hayes, R.J. and Zaba, B. (2001) 'Impact of mobility on HIV spread in rural East Africa', paper presented at IUSSP Conference, Session S28, Mobility and HIV in East Africa, Brazil, August.

Wiebel, W. and Safika (n.d.) 'Migration patterns of sex workers in Irian Jaya', Indonesia. PATH, Indonesia, Mimeo.

Wilson, M.E. (1995) 'Travel and the emergence of infectious diseases', *Emerging Infectious Diseases*, 1, 2: 39–46.

Yang, X.S. (2002a) 'Temporary migration and the spread of STDs/HIV: is there a link?', paper presented at the Association of American Geographers National Meeting, Los Angeles, March.

Yang, X.S. (2002b) 'Migration, socioeconomic milieu, migrants' HIV risk-taking behavior: a theoretical framework', paper presented at the Population Association National Meetings, Atlanta, Georgia, May.

3 Market bound

Relocation and disjunction in Northwest Laos

Chris Lyttleton

Introduction

Migration is by definition unsettling. People typically migrate in the hope of attaining something better, whether short or long term. This goal can require deliberate sacrifice of citizenship and social familiarity. Shortfalls in material and social entailments necessary for health often become apparent when people move to seek work. Transnational labour markets that promote large-scale movement are constituted by discursive formations that cut across political, cultural and economic realms. The labour migrant experience is a direct product of state controls that 'treat immigrant populations as the object of discursive elaboration, normalization and discipline and transform them into governable subjects as well as [labouring] ones' (Nonini 2002: 15). This transformation is not without costs felt intimately by these 'docile' bodies whose negotiable health can become the bargaining chip to attain access to labour markets.

Despite attention on migrant experience and diasporic processes, uneasy adoption of free market philosophies by many 'Third World' countries shows that systemics of global capitalism do not require physical movement *per se* to incur deeply felt changes in the social and individual body. Even as global circulation, which links time, space and labour in complex ways, has replaced capital penetration as the explanatory trope underlying latter-day modernization, how people and cultures are remade in the process remains the pressing question (Tsing 2000). This necessitates a focus on antecedent factors that motivate movement, as well as subsequent outcomes.

Social change that requires radical adjustment is also unsettling. Social change occurs wherever sociality is reproduced, that is to say everywhere; however, my interest in this chapter is when physical movement is combined with social policies that target specific practices. Most specifically, I focus on marginalization induced by forces of governmentality targeted at Akha highlanders in Northwest Lao PDR (henceforth Laos), one of a number of ethnic minority groups who are currently undergoing both relocation and rapid social change. Marginalization takes place when certain people and ideas are privileged over others. It shrouds minority groups such as the Akha

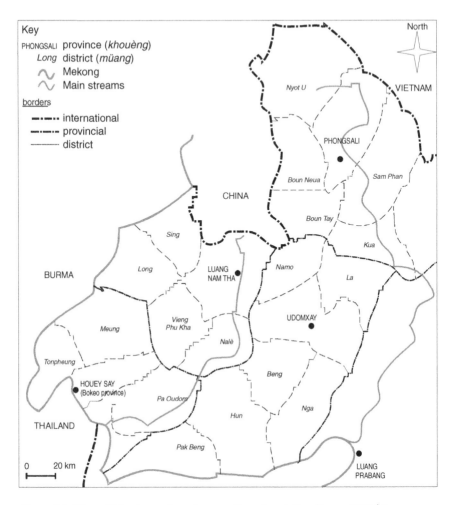

Figure 3.1 Map of Northwest Laos (source: Evrard and Goudineau (2004)).

when the bounds of normative behaviour shift; when socio-political and moral bounds, into which people within these groups are increasingly assimilated, become redefined through evolving and more tightly regulated social values. This reconfiguration has specific health consequences for those caught up in increasingly prosecuted regimes of knowledge and power that underlie nation-state formation.

In the almost universal embrace of capitalist modernity(s), local subjectivities are transformed by generic (albeit contested) processes that include growth of instrumental rationality; rise of national sentiment and nation-state; racialized perceptions of identity; spread of market relations and wage labour; birth of consumer societies; and an individualistic sensibility (Scott 2002: 167).

These characteristics may be observed in numerous states that seek to incorporate previously (semi-) autonomous ethnic groups living in peripheral zones somewhat beyond geographic and/or cultural control of centralized power structures. Their formation is particularly relevant in developing countries that seek foreign assistance to bring about national development in five-year chunks which conform to rationalized criteria of progress.

Nagengast (1994: 109) reminds us that national identity requires state-orchestrated pressure towards conformity, homogenization and assimilation, and societal agreement about what kinds of people fit (and what to do about those who do not). In the process of determining legitimate forms of social identity, assimilation also entails a politics of exclusion based on structural distinction and deprivation. Thus, Nagengast argues, in the process of establishing national identity, violence is often required. Subtle and discreet or blunt and abrasive, violence ensures that social and material conditions are embodied within a changing moral economy underpinning the creation and maintenance of collective nationalism. Violence is also implicit in targeted development projects seeking to modernize traditional societies through deeper engagement in free market economies. Rather than overt physical subjugation this is, as Derrida has argued, more commonly an exclusionary practice where one worldview displaces or minoritizes another within 'an economy of violence' (Parfitt 2002: 92). Changes in either instance are seldom absolute – works in progress – but the imprint on individual bodies is hardly transitory.

Recently cultural theorists have derived the term 'social suffering' to signal a conceptual approach that they believe is necessary to adequately understand the phenomenology of social experiences and rapid social change. Put simply, social suffering 'results from what political, economic and institutional power does to people, and reciprocally from how these forms of power themselves influence responses to social problems' (Kleinman *et al.* 1996: xi). In this chapter I consider how migration works in conjunction with mandated forms of modernization to jointly refashion life-styles and experiences of the Akha in Laos.[1] I wish to draw attention to how forms of social suffering emerge from 'everyday violence' as the Akha absorb various characteristics of modernity.

In the northwestern province of Luang Namtha, highlanders are moving down out of the mountains. They do not relocate far, sometimes within 20 km. But this is still distance of the early modern variety – measured by roughly two days' walk on difficult trails – and is not yet typified by the space–time compression that characterizes the 'postmodern' world and its movements almost everywhere else. Regardless of whether local villagers choose to relocate closer to roads and markets, agree to state pressures that say they would be better removed from 'threatened' forests, or decide they will stay put come what may, social change is rapidly altering daily life. Subsistence pursuits are quickly giving way to a cash economy, and insofar as modernity inevitably requires periodization (Jameson 2002: 29), we suggest, with no intention of being trite, that times are changing for the Akha of Luang Namtha.

In recent years socialist Laos has cautiously engaged in a free market economy and global trade. In mountainous regions throughout the country geographic movement combines with a temporal shift as ethnic villagers who reside in these remote areas are called upon to embrace the 'new', defined prominently by 'increased flow' (Tsing 2000: 345). For local highlanders, flows come in a number of distinct forms marked by greater exchange of ideas, money and goods. New mandates arrive via state and donor agencies on a regular basis, and new roads bring people, materials and ideas from other provinces, countries and cultures. Signalled by recent targeting of 'pro-poor' projects, capitalist modernity has arrived in Luang Namtha, and with it the confrontation of traditions that has played itself out in numerous indigenous and peasant communities the world over. Timeless traditionalists (as they are often romanticized) no longer, highlanders are being made over as citizens of 'modern-day' Laos.

Although mobility has been commonplace in the highlands, contemporary migration downhill, even over short distances, requires significant cultural change and adaptation. In this instance, transformation takes its cue in a radical shift required in villagers' life-styles. The Akha are told that they must stop slash-and-burn and they can no longer grow opium. With these specific mandates, change is not subtle or gradual (although it also has these characteristics). It is extreme and seemingly non-negotiable. Life in the mountains, as it stands, has suddenly become unsustainable, not through ever-present hardship, which is hardly unfamiliar, but through a new set of ground rules that dramatically reconfigure everyday livelihood. The common response to date requires movement – away from forests, away from subsistence and away from life-styles based primarily on local exchange. And in the course of this movement an important boundary is crossed; not a national border, but a boundary defined by elevation. The shift from highland to lowland requires greater transformation than that implied by physical distance alone, and this is what makes the term *migration* appropriate. In contrast to prior modes of mobility, it entails active renunciation of key characteristics that have become pejoratively associated with highland life – forest-destroying, drug-producers – and the embrace of lowland values based on market enterprise, consumerism and wage labour. Needless to say, Akha villagers embody these changes in ways that they do not always control or anticipate. Health is therefore an obvious optic one can employ to consider the subjective and embodied impact of mobility and imposition of modernity.

In Laos, health is understandably the focus of numerous development initiatives. The population is relatively young; almost 55 per cent of 5.5 million are under nineteen years – only 4 per cent over sixty-five years. Health standards are low by any international standard and the worst in South-East Asia: the overall crude mortality rate is 15/1,000; maternal morbidity is 656/10,000 and infant mortality 104/1,000. Life expectancy has recently increased to fifty-two years for women and fifty-nine years for men. Each of these indicators is notably worse among highland groups where language barriers, severe poverty, malnutrition, illiteracy, superstition, non-hygienic life-styles and opium growing are regarded as persisting obstacles (Government of Laos PDR 2003: 83). In addition, and while many donor-based projects are attempting to improve this situation, what receives less attention is how modernization is itself creating vulnerabilities to ill-health.

The Akha, resettlement and marginality

The Akha, of the Tibeto-Burman language group, are no strangers to movement. Although census data are unreliable, the Akha number approximately 0.5 million (Kammerer 2000). They have gradually moved southwards for several hundred years and live throughout the mountainous upper Mekong: in Southern China (150,000), Burma (180,000), Northern Thailand (33,000) and Vietnam (12,000). Nearly half of the 60,000 Akha who arrived in Laos over the past 150 years now reside in Luang Namtha Province, primarily in two districts: Muang Sing and Muang Long.

Prior to programmes to integrate them into centralized forms of state governance, Akha, along with other ethnic groups, inhabited elevated forest zones practising subsistence swidden economy and hunter-gathering. Thus, the common (but misleading) label attached to highland minority groups as being 'semi-nomadic' highlights long-term familiarity with movement and creates the opportunity for accusations of illegal land occupancy (Kammerer 1988). In China and Thailand, initiatives to include ethnic minorities within national programmes covering both social and environmental management have been in place for decades and have all but stopped such movement. In Laos, national policy directives to halt swidden agriculture have a more recent history.

While the Lao government has been concerned with controlling shifting cultivation since it took power in 1975, progress has been haphazard at best, in part due to local and foreign-donor resistance to potential coercion in 'focal site' resettlement initiatives (UNDP 2001: 4.3.2). Nevertheless, the Lao government is increasingly insistent – the Seventh Party Congress in 2001 declared a policy that would 'eliminate opium production by 2005 and put an end to pioneering slash and burn cultivation by 2010' (Government of Laos PDR 2003: 8). It is the implementation of this combined objective that is now causing major upheaval for Akha in Luang Namtha.

The Lao Akha are but few of an estimated 30 to 80 million people who depend on shifting cultivation in the Asia-Pacific region, most frequently ethnic minority groups (FAO 1993: 12). Throughout the region, processes of assimilation and state attempts to control swidden foster tension and conflict across ethnic lines. Thus present-day circumstances in Luang Namtha are not atypical. In numerous historical examples, and still today, violence commonly accompanies changing parameters of livelihood choices. In Burma, ethnic insurgency and high mortality is commonplace (Feingold 2000); in Vietnam, recent ethnic minority uprisings have been linked to land encroachment (*The Economist*, 7 April 2001); in Cambodia, armed confrontations take place between loggers and highlanders (Bottomley 2002); and in Thailand, highlanders have been punished with murder, enforced expulsion and a litany of hardships for supposed illegal practices associated with swidden (Chupinit 1994). Movement and migration are a frequent response to these hostilities and the dislocation has a marked impact on everyday health status of ethnic highlanders.

But not all movement is coerced – nor all violence enacted with overt force. Kleinman (2000: 226) has argued that everyday violence occurs when vectors of economic and political power incur damage to physical body and moral experience. In a similar vein, Thai critic and Buddhist commentator Sulak Sivaraksa (2002: 47–9) contends that 'quality of life' is diminished by 'social and environmental violence' implicit in value systems underpinning neo-liberal capitalism that place impersonal profits ahead of inclusionary concern for social diversity and the environment. One must be careful not to overstate the case. Liberal capitalism has undeniably fostered tremendous improvements in living standards the world over. But for every claim of its success there is a counter-claim of capitalism's depredations on human and environmental well-being, and ample evidence of those who miss out on benefits attributed to 'the magic of markets' (*The Economist*, 28 June 2003: 15). Careful attention to specifics is necessary before such adjudications can be made, as we are ill-equipped to measure and categorize the social adversity and physical and emotional pain brought about by structural deprivation. In this sense, Sulak's notion that capitalism excludes aspects of human experience is apposite. A UNDP assessment of community marginalization in Thailand emphasizes the disempowerment emerging from decades of development:

> Communities have no wish to reject modernity, oppose globalization, and cling to the past. But they want power to determine the direction of development based on their own body of knowledge, their own values, the principle of sustainable balance between man and nature, and the community's rights to manage resources.
>
> (UNDP 2003: 79)

Within this framework of uneven power relations, the halting of swidden agriculture (and opium eradication) has become fundamental to minority group assimilation within the upper Mekong. A politics of ethnic difference, premised on threats that traditional minority practices pose to the environment and national security through migration, rebellion and illicit drug production, has often promoted punitive and discriminatory action from dominant (ethnic) cultures (McCaskill 1997). Significantly, it is not only national borders that figure in the nationalist reckoning of the 'minority problem'. So too elevation becomes a defining characteristic of a spatial zone that receives increased state surveillance and policing (Feingold 2000). Highlander traditions, criminalized by national policies, justify official intervention including fines, arrests and resettlement. State forces demand more than an enforced surrender of access to forest resources. Socio-economic changes under the guise of instilling new systems of food and cash-crop production inevitably intrude within complex realms of ethnic identification and cultural value.

It is not simply that minority groups seldom negotiate these social and cultural changes from a position of advantage. Also at stake is the sustainability of cultural integrity itself (Choosit 2003: 78). As Henin notes, the Akha, 'faced with loss of land and deterioration of their environment, are in a continuous process of adapting to outside forces in order to survive as a cultural entity' (1996: 180). In turn, the introduction of new value systems has widespread implications for the health of minority populations. The Akha in Thailand have been subject to severe cultural disjunctions following 'development' initiatives that result in large-scale movement to towns and cities seeking wage labour, widespread uptake of heroin, commonplace female prostitution and devastating epidemics of HIV/AIDS and malnutrition (Geusau 1992; Toyota 1996).

Similar state and market-driven forces that have precipitated change in other parts of the upper Mekong are now taking place throughout Laos, in particular the prohibitions on swidden and opium production. Issues of ethnic marginalization are pan-regional but the political and economic specifics vary in significant ways. Thailand, for example, has a legal apparatus that places highly bureaucratic and militant scrutiny on ethnic groups populating the mountain areas and their ability (or lack of) to claim Thai citizenship. In contrast, Lao national policy (since the communist revolution) has staunchly advocated ethnic equality, and swidden control is facilitated by policies that encourage highlanders to relocate and become active members of a Lao citizenry engaging in a free market economy, thus effectively extending forms of governmentality and control (Evrard and Goudineau 2004).

In Thailand, orchestrated relocation out of the hills has rarely proved successful, and forms of nationalist control are more typically marked by the opposite movement into the highlands of commercial interests accompanied by legal strictures on forest protection, citizenship and national security. By contrast, as the Lao highlanders enter lowland styles of production, they more commonly negotiate rather than contest their positions within a new capitalist order. In some parts of the country, armed violence between state forces and ethnic groups (such as the Hmong) occasionally flares, but, rather than prompted by relocation projects *per se*, this is generally explained in terms of long-standing tensions etched during the Vietnam War. While limited resources prevent many development promises from

being readily achieved, Lao ethnic policies have lessened (although not erad-
icated) the scale of official (and sometimes violent) discrimination which
minorities in other parts of the region still face. None the less, insofar as
modernity inevitably produces exclusion and hierarchy (Giddens 1991: 6),
processes of assimilation have a noticeable impact on everyday livelihood in
Muang Sing.

Movement and disease in Muang Sing

In 2003, Muang Sing comprised ninety-seven villages and 29,307 inhabitants. Akha make up approximately 45 per cent of the district population in
sixty villages. In the past, these villages were typically established in the
mid-upper slopes surrounding Sing valley. For the past twenty years, there
has been a gradual but steady movement of Akha to the lower slopes and
valley surrounding Muang Sing town.

By 1995, thirty-five villages had been recently established in the lower
slopes and about twenty-three remained in the highlands (Gebert 1995).
The lower slope villages co-reside with other ethnic groups such as Tai Lue,
Tai Dam and Tai Nua who have historically cultivated paddy-rice in valleys
throughout the region. Relocated villages form as subgroups of families
splinter from older villages – a scission that has long characterized Akha
social life and the need for continued access to productive land. Movement
to the lowlands has one significant difference – a shift in rice-growing techniques from rotational production to sedentary paddy-field cultivation. Its
adoption signals a willingness to embrace 'lowland culture', and movement
to establish permanent lowland villages has often been led by younger Akha
(Cohen 2000). Networks typically remained strong between parent–
offspring villages and the lowland communities have gradually grown
bigger through subsequent kin migrations. Many lowland villages now have
two distinct household clusters marked by older and newer arrivals, with
varying rights and access to productive land.

One point needs to be clearly stated. In the 1990s when the bulk of this
relocation took place, movement was not the simple product of state mandates but rather a combination of push-and-pull factors. Although Gebert
(1995: 11) notes that 'many of these villages have moved down to lower
slope areas at request of government authorities and were also promised
assistance if they would be ready to cultivate flat land and abandon shifting
cultivation practices', Cohen (2000: 185) suggests motivations also include
low productivity of swidden agriculture (rice and opium) in the highlands.
Despite an uncertain future and a departure from traditional life-styles,
many of the current generation of Akha consider improved market
opportunities, proximity to transport routes and social changes implicit in
becoming lowlanders to be a challenge worth taking.

However, one clear loss is evident – many people die. Total numbers have
not been recorded but lowland villages report large numbers of deaths

during the first two years of resettlement. Headmen describe between fifty and 100 villagers dying shortly after relocation – sometimes whole households perish. With data derived from the interviews of over 150 families, Gebert (1995) compared health levels in eleven mid-upper slope villages and twenty-two relocated lower slope villages, as seen in Table 3.1.

In the neighbouring district of Long, a 2003 survey of mortality over the preceding five years confirmed high death rates, sometimes up to 20 per cent (Romagny and Davineau 2003: 21), as seen in Table 3.2.

While mortality is high in the highlands, movement to the lower slopes further increases death rates. High mortality stems from a series of inter-related factors: lack of physiological immunity to mosquito-born diseases; lack of access to clean water or familiarity with sanitation; and a lack of village-based health services and disinclination to access state medical assistance. Combined, these shortfalls have led to high levels of misery and, until recently, an ongoing reliance on opium to cope with psycho-social trauma (Cohen 2000; Gebert and Chupinit 1997).

Despite high death rates, Akha continue to relocate. Noticeably, over the past eight years some health indicators have gradually improved in Muang Sing. In Deutsche Gesellschaft fur Technische Zusammenarbeit (GTZ) target villages, infectious diarrhoea and parasite-borne disease decreased from 29 per cent in 1996 to 13 per cent in 2000, and under-5s mortality decreased from 311/1,000 in 1997 to 105/1,000 in 2001. But deaths related to relocation still occur: in 2003 one village had nine deaths within a group of 116 migrants over a period of eight months – a rate of 7.8 per cent (Alton and Houmpanh 2004: 46).

Table 3.1 Comparison of disease and child mortality between highland and relocated villages in Muang Sing

	Mid-slope (11 villages)	*Lower-slope* (22 villages)
Disease epidemics	6	24
Epidemic deaths	199	749
Child Mortality (per 1,000 live births)	133	326

Source: Gebert (1995).

Table 3.2 Comparison of death rates between highland, lowland and relocated villages in Muang Long

	Average mortality (per 100)
Existing lowland villages (n=5)	0.78
Highland villages (n=15)	2.32
Resettled villages (n=17)	3.99

Source: Romagny and Davineau (2003).

Some improvements in morbidity levels notwithstanding, it would be wrong to say that highland–lowland migration has become less problematic. Crucially, in this context, health is not simply about the absence or control of recognized disease vectors. The damaging impact of social change should not only be measured by bodies after they become ill or die. There are other variables that foreshadow the emergence of social pathologies and influence subjective states of well-being. In this context, social suffering, based on changing social relations and economic structures, plays its part in predisposing individuals to subsequent conditions of disease. While it is not possible to empirically measure levels of social suffering, the effects are anything but ephemeral. Importantly, this perspective draws attention to social relations, rather than material conditions, as the underlying preconditions for ill-health.

In Luang Namtha, conditions underpinning new forms of ill-health crystallize in the increased pressure to migrate out of the hills. In most previous village migrations down to the lowlands the Akha have maintained strong links with the parent villages through regular exchange of household commodities including opium. Recently, however, a new imperative has caused the Akha to move more rapidly and precipitously and has reduced their ability to control incorporation into the lowland economy. The fraught achievement of radically new social competencies required by sedentary settlement, and reliance on wage labour, is creating vulnerability to a range of public health threats including HIV and changing forms of drug abuse.

In early 2003, following a Ministerial decree, officials ordered the destruction of a large percentage of poppy fields. Previously, the Lao government had shown little interest in strong-arm opium eradication. Pressure to tackle deforestation and appease foreign donors interested in drug reduction has altered the lenient approach. Opium is a vital cash-crop for highland Akha throughout Luang Namtha. Apart from sustaining high levels of local addiction and a range of medical and social needs, opium is an important exchange item for household necessities such as salt and utensils and a crucial buffer against rice shortages. In 2003, with no opium to trade, there have been immediate rice shortages in many highland villages.

In contrast to previous movement based on persuasion and enticement, relocation is now a product of direct intervention. District officials in Sing and Long recently announced that nearly 50 per cent of highland villages were slated to be moved or consolidated. In a short space of time, Akha in higher slopes have learned that choices that had been theirs to make for preceding generations were, according to state fiat, no longer tenable. In 2002/2003, movement out of Muang Sing hills snowballed. Subsequent confusion is aptly summarized in the following report:

> It is estimated that about 15 villages with about 2,000 people from the mountains moved to lowland areas because their poppy fields were cleared. 'We knew before the clearing of the poppy fields that villagers

would move', the District Vice Governor informed the team. 'On the one hand, it was good that they moved. For many years we had asked them to do so, but they did not. However, on the other hand, it made things more complicated because the district could not carry out the development as planned. Villages were messed up everywhere, it was not according to our plan'. 'We have to stop all development activities because of migration: everyday villagers ask the district authorities to find a new place for them to live', the Vice Governor added.

(GTZ 2003: 15)

Negotiations with kin for co-residence in existing villages or officially approved acquisition of new land can take months. Obtaining productive land is not guaranteed. In the meantime, villagers, adamant over their need to relocate due to economic hardship, live in what look like hastily constructed refugee camps on land which the state designates (at times would-be migrants are turned back to their home villages, but seldom successfully). Even when highland villages have been reluctant to move, the exodus of neighbouring villages has made it essential due to lack of proximate trade and exchange networks. Some higher slope areas are now completely abandoned.

In the light of my concern with 'everyday violence' emergent in situations of drastic social change, the question is: What impact does this movement, voluntary or forced by circumstance, have on the lives of the Akha in Muang Sing? It remains to be seen whether large-scale malaria epidemics and waterborne diseases recur. But more subtle forms of disease are also established as a product of these changes. They apply not only to recent arrivals, but also to processes of social and cultural assimilation that have been underway for the past ten to fifteen years in Muang Sing (and much longer in neighbouring countries) in the transition of the Akha to subjects of a modern nation-state.

Embodying change

Numerous studies show how individual and social bodies cope (or do not cope) with cultural confrontation under projects of colonialism and/or development. Sometimes, extreme physical suffering results from structural violence and disruption of local value systems. Taussig (1984) provides harrowing details of violence in the imposition of radically alien modes of (capitalist) production and forced migratory labour in colonial rubber plantations in Putamayo lowlands, Columbia. Not all forced entrances of pre-capitalist societies into the market economy, and its attendant structures of class hierarchy and moral hegemony, have been marked by the culture of cruelty and fear, but there is no disputing the common distress in the confrontation of 'difference'. But a phenomenology of suffering associated with social change encompasses more than brutal exploitation. Inequitable forms of governance and structural violence also promote insidious health dilemmas, such as HIV spread (Farmer 1996). As such, the presence of illness frequently links to existing structures of marginalization affecting women and ethnic groups the world over. Sugar, Kleinman and Heggenhougen have argued that modernization is implicated in widespread rise of social and psychological pathologies such as depression, suicide, substance abuse and psychoses through what they term 'noxious change': 'Psychological and social distress appear to be increasing worldwide as a direct effect of certain social changes, including those that are "planned" in international-development projects' (1991: 217).

Despite a historical backdrop of regional rebellions, revolutions and forced relocations, Akha in Luang Namtha have largely been left to their own devices in recent decades. In the absence of roads, remoteness has been the overriding characteristic. It is therefore inappropriate to talk of overt violence affecting their health through government policies or donor-driven development programmes. But if we accept that marginalization is implicit in processes of modernization, then social suffering becomes an additional dimension necessary for examining the lives of minoritized peoples and their experiences. However, marginalization is a phenomenon seldom adequately explored for its health implications in projects of state assimilation. For, even as public health assistance has somewhat reduced quantifiable levels of morbidity in developing countries, degrees of social suffering are far harder to grasp through empirical measures.

In recent years, the lives of the Akha have been recorded in fine detail: family statistics, and hours spent hunting, gathering, crop-growing, firewood-collecting and opium-smoking are collated and analysed. However, despite numerous finely tailored initiatives by donor agencies, it is regional stereotypes of Akha as a forest-destroying, drug-producing, semi-nomadic, undeveloped ethnic minority that have defined the most forceful interventions that are now brought to bear. Thus state-driven land allocation and zoning strategies are being carried out that will limit Akha access to forest resources. Opium fields have been laboriously counted and measured and largely destroyed. Anti-HIV workshops attempt to drastically restructure adolescent courting patterns.

Such imperatives, including new ecotourism initiatives, are part of the so-called 'pro-poor' agenda of state and multilateral development agencies. As I have been describing, the end result for many Lao Akha, designated as the poorest of the poor, is relocation into the lowlands. Ironically, in a government analysis of poverty (based on data from thirteen provinces), relocation and reconsolidation are cited as key variables associated with increased poverty: 'Agricultural production, however, has not improved and has often become worse especially after village relocations, and is not sufficient for sustainable livelihoods' (Government of Laos PDR 2003: 45). A recent study confirmed that livelihood systems suffer and health impacts are commonplace in relocated villages in Sing and Long Districts: 'The main problem encountered with migration to the lowlands concerned inadequate available agricultural land and this led to conflicts with neighboring villages and earlier migrants. Another major problem was the incidence of human diseases' (Alton and Houmpanh 2004: 14).

Yet despite widespread negative consequences – devastating epidemics, loss of assets, rapid debt accumulation, reduced rice supplies, intensified competition for land and lack of government resources to provide assistance (Goudineau 1997) – the policy emphasis on access to markets and roads and the sedentary production of cash-crops is based on the opposing notion that livelihood improvement will take place automatically if villagers move to the lowlands. As mentioned above, this assumption dovetails with the present-day aspirations of many Akha. Several male informants in recently relocated villages noted that despite the gruelling aspects of day-to-day survival in the baked land of newly cleared encampments, simple recollection of back-breaking highland life halted any reticence. Not all women agree, but overall distaste for the highland livelihood and its lack of material symbols of development figured strongly in the desire to move. In marked contrast, lowland life-styles are believed to facilitate the accumulation of overt indicators of progress, such as 'modern'-style houses and commodity goods, in particular motorcycles or tractors. Coupled with the removal of opium as the solitary source of wealth in the mountains, the desire for accumulated symbols of material well-being adds to the forces motivating movement (some villagers are, however, beginning to question the value of relocation, and occasionally, beckoned by promises from Chinese rubber investors, have returned to the highlands).

At what point, then, does social suffering become implicit in migration impelled by both government initiatives and prevalent desire for capital and material improvement, given the fairly radical changes towards a proletarian life-style necessary to achieve these objectives? In the face of historical examples throughout the region that highlight marginalization and ill-health underpinning assimilation, the issue requires attention. Clearly it cannot be assumed a priori that rapid movement into a market economy is detrimental to health. But if, on the one hand, poverty is associated with relocation and, on the other, poverty is a major risk factor for most forms of social suffering (Kleinman *et al.* 1996), consideration of health impacts that emerge as a product of transformed aspirations is warranted. Closer inspection does indeed determine levels of vulnerability not yet quantifiable in epidemiological terms, but of major importance to future health.

Social suffering in Muang Sing

Poverty and exploitation are clear consequences of relocation. Despite dreams of wealth, purchasing and developing productive land requires substantial de-capitalization – a burden many families never overcome. While some families might improve economically, this is a notable minority wherever relocation takes place: 'In the villages where overall income had increased, it was found that increases in production were being realized by only a few families' (Government of Laos PDR 2003: 45). Likewise, Alton and Houmpanh (2004: 44) highlight diverse problems in Sing and Long – all but one of seven study villages had rice deficits due to inadequate land availability.

A majority of relocated villagers, therefore, become dependent on full-time, off-farm wage labour. Although access to labour is the ostensible appeal of relocation, increased reliance begins a spiralling process of prole-tarianization. Rice shortages ensure that villagers have to labour for other, wealthier villagers or neighbouring ethnic groups, so that even if land is available it is never developed into productive rice-paddy. Cohen showed that in two Akha villages most households were worse off after relocation due to disadvantageous labour relations with other ethnic groups (2000: 197). In the past, Akha may have occasionally laboured for cash, but very few were totally reliant on it because they had access to opium and abundant forest resources for food and trade. While conflict over land is commonplace in contexts of relocation, it is the health implications of attendant proletari-anisation that I wish to highlight.

First, exploitation and conflict escalate as a consequence of obligatory wage labour. In contrast to primarily subsistence livelihoods, waged employment allows articulation of uneven power relations which can become a key precursor of social suffering. In the lower slopes waged employment takes place most commonly with neighbouring ethnic groups such as the Tai Lue (who have historically hired Akha addicts for minimal cash or opium) or Chinese market gardeners who invest in sugar and water-melon production. Near the border, Akha women cross into China to work in the paddy-fields there.

With increased migration, wage labour has become a surplus commodity in the Sing lowlands and increased competition is emerging within villages. On the one hand, neighbouring ethnic groups employing casual daily labour have begun to hire only Akha between the ages of twenty and thirty-five, severely disadvantaging many older householders. On the other hand, the need to be actively seeking work is unevenly advantaging subgroups within relocated villages. Some village women told us that they were too shy to repeatedly go out asking for work in nearby villages when they had been turned down once, and yet this daily demand for labour is the precise result of current employment arrangements. Other subgroups (stemming from their arrival from different source villages) were seemingly not so reticent, thereby attaining the bulk of local work available and creating the begin-nings of village factionalism. Unused to competing for a wage, villagers described resentment born of this biased distribution of work. It is further exacerbated when the demand for labour is limited and certain kin groups (again based on different home villages prior to migration) will only mobi-lize work-gangs from those within their subgroup, setting in place the gradual but insidious advantages that accrue from greater access to capital.

Labour exploitation also takes place within villages constituted by sequential waves of migration. Wealthier villagers hire newcomers who are in need of instant cash to work their fields, making it impossible (even if land is available) for them to invest time and labour to become self-sufficient. Addicts have been particularly susceptible, since they exchange labour for opium and inexorably incur increasing debt. As a result, conflicts over land and labour exploitation between older settlers and newer arrivals is becoming characteristic within relocated villages, and communal activities, the hallmark of Akha society, are fractured and at times totally dysfunctional. At other times, unscrupulous Akha brokers or outside traders have refused to deliver promised wages from the sales of locally harvested sugar to Chinese factories.

Competition is the heart of market-based capitalism. In Muang Sing, growing labour competition is setting in place the basis for entrenched village factionalism, requiring new styles of conflict management and adjudication than those typically employed in the past. While village conflicts and degrees of social stratification have always existed, they have seldom been based on struggles over sale of labour. Previously, competition over resources was often resolved by one faction moving to establish domicile elsewhere in the forested hills. In the lowlands, where agricultural land is now a scarce commodity and forest products a diminishing resource, conflict avoidance in this fashion is no longer an option. The stage is set for increased deprivation among some villagers who are unable to compete successfully for labour and, at the same time, escalating village conflicts based on the inability to negotiate more equitable social and economic relations.

Second, new forms of drug abuse are also creating psycho-social problems whose costs are still being counted. Drugs have long been an integral element of everyday Akha livelihoods and, in new forms, continue to play an important role in health dilemmas. Opium has played a historically complex part in local economies throughout the region and, until recently, levels of addiction among the Akha were high (9.8 per cent in Sing and Long districts). Drug rehabilitation has therefore been a basic component of most development initiatives in the area. As I, and Paul Cohen, have explored elsewhere (Lyttleton and Cohen 2003; Cohen and Lyttleton 2002), demand reduction is problematic for most addicts, many of whom relapse, exacerbating social hardship and marginalization. The recent upsurge of amphetamine-type substances (ATS) throughout South-East Asia creates additional problems.

ATS was introduced several years ago by petty traders within enclaves of townspeople in Sing and Long districts, but its consumption is moving steadily outwards into Akha villages, primarily those in proximity to the town or road. Data collected by district officials in 2003 identified 800 local ATS users throughout Sing District, of which approximately 30 per cent were Akha. Now that opium cultivation is stringently policed, local Akha addicts are using ATS (rather than heroin which was taken up in Thailand) as a means of decreasing opium reliance. Significantly it is not only opium addicts who smoke ATS; labourers in a wide range of contexts use it to facilitate their entry into new market economy structures.

I have argued elsewhere that social values associated with modernization predispose people to drugs that foster improved performance (Lyttleton 2004). ATS use is growing rapidly throughout South-East Asia in part as a means to improve labour productivity, and from this perspective, increased ATS use is logical for Akha entering new social and economic relations in Sing and Long valleys. In addition to providing overt energy, ATS use facilitates desirable 'modern' forms of subjectivity wherein traits associated with ethnic traditions, such as opium addiction, lethargy and 'primitiveness', are exchanged for an active entrepreneurial labourer/trader identity. As opium addiction is gradually removed, incipient ATS abuse poses growing dilemmas for those attempting to benefit from its use. This relates both to individual health and social pathologies. Prolonged use creates dangerous forms of psychosis and paranoia. The number of arrests for ATS possession and use is increasing in Muang Sing and the number of violent crimes associated with its use is rising, both within Akha and other ethnic groups.

Third, desire for cash and material goods implicit in movement out of the mountains heightens other forms of disease vulnerability, in particular HIV. Adoption of lowland life-styles is bringing diverse changes within Akha villages. Prominent indicators of expanding 'globalized' interactions include the bars and nightclubs proliferating along the new road that bisects Muang Sing, and links China with the Mekong River and Thailand downstream. Chinese trucks move back and forth in convoys. Akha men are starting to make exploratory use of 'hospitality services' provided by itinerant women from other parts of Laos and China in bars being built in close proximity to relocated villages. More significantly, local Akha women are beginning to exchange sex for money. In Akha villages, following traditional customs and negotiations, young women will sometimes sleep with visitors to the village. Nowadays, in many relocated roadside villages, Chinese migrant labourers and Lao men from towns arrive some evenings to attempt to negotiate sex with local women. Whereas in the past, in highland communities, whisky or cigarettes for village youth were accepted coinage, in more established lowland villages young women are increasingly expecting to be paid directly for such services, reflecting the ascendancy of cash as the currency of exchange.

The combination of a growing sexual opportunism of migrant Chinese labourers and local Lao men and incipient forms of sexual commercialism within Akha communities are huge threats to health and the cultural future of minority groups in the area if HIV is part of these exchanges. So far no testing has been done to ascertain this (nor any effective campaigns to alert Akha women and men of potential risk), but already endemic levels of gonorrhoea in some villages are a testament to rapid STD spread. The intersection of short-term migrants using newly built roads and highlanders moving to be near markets and services poses serious health threats. When actual numbers of individuals begin testing HIV positive, much damage will have already been done.

Conclusion

Modernity as the key narrative of contemporary nation-states requires specific social, political and economic formations. Nationalized permutations of modernity's demands have homologies in the bodies of their subjects. These are, of course, not uniform across a state's citizenry. Forms of exclusion and structural deprivation that gird a nation-state's growth ensure that privileged and disadvantaged bodies are impacted in a divergent manner. According to Arrighi (2002: 42), capitalism as an intrusive systemic of rule and accumulation is a 'key interstitial formation of both pre-modern and modern times'. The highland minorities of the upper Mekong embody this interstitial condition in awkward and, at times, notably damaging, ways.

In Thailand, highlander assimilation has not incurred the pace of current movement to lowlands that we see in parts of Lao PDR with its doctrinal policy of universal citizenship, equal rights and less strident competition for land. Instead, commercial interests moving upwards into Thai highlands have restricted hilltribe access to resources and dislocated them from traditional life-styles leading to disease and 'a growing sense of despair' (Kammerer 2000: 47). In Southern China, the introduction of commercial farming has seemingly had more positive impacts on the Akha than in neighbouring countries with an increase in living standards through tea production. But there are also growing inequities within and among Chinese Akha villages and pronounced cultural domination by national policies (Henin 1996).

In Laos, migration and assimilation are a complicated mix of push-and-pull factors: of state pressure and individual desire for access to money and goods. So too health impacts are diverse. On the one hand, government and foreign programmes have somewhat reduced common endemic infectious diseases among the Akha. On the other hand, health is now more closely determined by motivation to engage in a cash economy, to exchange physical risks of impersonal and fragile subsistence ecology with broader social and psychological risks of an impersonal and high-stakes market. The lowland valleys of Sing and Long are currently home to a highly diverse yet intensive mix of social and economic aspirations. It is within this crucible of change, opportunity and exploitation that people's lives are shaped: physically and psychologically; socially and materially. The Akha in Muang Sing have in the past fought for their health largely against an unforgiving physical environment, and high mortality rates attest to harsh terms of negotiation. If infectious epidemics such as malaria and water-borne disease are controlled, then it becomes the social environment where relocated Akha negotiate a new set of defining rules and values that poses a greater threat to health. Here the uneasy intersection of different cultural systems promotes inequitable economic and social interactions that foster social suffering through marginalization and exclusion. Although not immediately visible, they create debilitating forms of distress.

Poverty is nothing new to the Akha, but, as social stratification is refined through limited access to wage labour, it will be experienced differently by relocated Akha. Factionalism and conflict over resources will increase as traditional resolutions such as mobility are less viable. Clearly not all will suffer, but relocation becomes disjuncture when the shift from pre-modern to modern induces new forms of enduring structural disadvantage and deprivation (including loss of that amorphous human right – dignity). For just as capitalism is the bridge to the modern, transitions have benefits and costs. One Akha man living by the road compared the change: 'In the mountains life was better, we had more money [from opium] but nothing to spend it on; here by the road we have less money, but more [desire for] things to buy.' Coupled with state pressures to move, desire to be part of a market economy is the fundamental coinage for the social changes taking place; it is the ticket to a new life and, at the same time, the harbinger of disease present and still to come. Nascent processes of marginalization, experienced as factionalism and community disempowerment at the group level, and exploitation and relative deprivation at the individual level, are inevitable underpinnings of disjunctions emerging from uneasy integration into new social and economic systems. In turn, they underlie forms of social rather than material suffering that emerge with the movement from small-scale (primarily) subsistent communities to sedentary groups partaking in larger networks predicated on a market economy. Unchecked, the everyday violence they reflect will have overt health consequences such as new forms of drug abuse and new forms of infectious disease such as HIV.

Note

1 This chapter is based on ethnographic fieldwork conducted over ten months between 2000 and 2003, in Sing District (and to a lesser extent Long District), Luang Namtha Province, Laos (see Figure 3.1). With an Akha research assistant, I interviewed community leaders, government officials and residents in a sample of twenty-five lower and mid-slope Akha villages, including several which had relocated in the previous twelve months. In 2003, this study formed part of a larger project conducted in collaboration with the Lao Institute for Research on Culture, supported by the Rockefeller Foundation and Macquarie University.

References

Alton, C. and Houmpanh Rattanavong (2004) 'Service delivery and resettlement: options for development planning', unpublished report, Lao PDR, Vientiane.

Arrighi, G. (2002) 'The rise of East Asia and the withering away of the interstate system', in C. Bartolovich and N. Lazarus (eds) *Marxism, Modernity and Postcolonial Studies,* Cambridge: Cambridge University Press, pp. 21–42.

Bottomley, R. (2002) 'Contested forests: an analysis of highlander response to logging, Ratanakiri Province, Northeast Cambodia', *Critical Asian Studies*, 34, 4: 587–606.

Choosit Choochat (2003) 'Indigenous wisdom, forest preservation and ecotourism in Northern Thailand', in A. Nozaki and C. Baker (eds) *Village Communities, States and Traders*, Bangkok: Sangsen Publishing House, pp. 68–83.

Chupinit Kesmanee (1994) 'Dubious development concepts in the Thai highlands: the *Chao Khao* in transition', *Law and Society Review*, 28, 3: 673–86.

Cohen. P. (2000) 'Resettlement, opium and labour dependence: Akha–Tai relations in Northern Laos', *Development and Change*, 31: 179–200.

Cohen, P. and Lyttleton, C. (2002) 'Opium reduction programmes, discourses of addiction and gender in northern Laos', *Sojourn*, 17, 1: 1–23.

Evrard, O. and Goudineau, Y. (2004) 'Planned resettlement, unexpected migrations and cultural trauma: the political management of rural mobility and interethnic relationships in Laos', *Development and Change*, 35, 5: 937–62.

FAO (1993) *Challenges in Upland Conservation*, Bangkok: Regional Office for Asia and Pacific.

Farmer, P. (1996) 'On suffering and structural violence: a view from below', *Daedalus*, 125, 1: 261–83.

Feingold, D. (2000) 'Kings, princes and mountaineers: ethnicity, opium and the state on the Burma border', paper presented at Beyond Borders Conference, Paris, July.

Gebert, R. (1995) 'Socio-economic baseline survey', unpublished report, Muang Sing: GTZ.

Gebert R. and Chupinit Kesmanee (1997) 'Drug abuse among highland minority groups in Thailand', in D. McCaskill and K. Kampe (eds) *Development or Domestication: Indigenous Peoples of Southeast Asia*, Chiang Mai: Silkworm Press, pp. 358–97.

Geusau, L. Alting Von (1992) 'The Akha: ten years later', *Pacific Viewpoint*, 33, 2: 178–84.

Giddens, A. (1991) *Modernity and Self-Identity: Self and Society in the Late-Modern Age*, Oxford: Polity Press.

Goudineau, Y. (1997) *Resettlement & Social Characteristics of New Villages: Basic Needs for Resettled Communities in Lao PDR*, Vientiane: UNDP.

Government of Laos PDR (2003) 'Lao PDR's national poverty eradication programme: a comprehensive approach to growth with equity', unpublished report, Vientiane.

GTZ (2003) 'Study report: drug free villages in Sing District, Luang Namtha Province', unpublished report, Muang Sing.

Henin, B. (1996) 'Ethnic minority integration in China: transformation of Akha society', *Journal of Contemporary Asia*, 26, 2: 180–200.

Jameson, F. (2002) *A Singular Modernity: Essay on the Ontology of the Present*, London: Verso Press.

Kammerer, C. (1988) 'Of labels and laws: Thailand's resettlement and repatriation policies', *Cultural Survival Quarterly*, 12, 4: 7–12.

Kammerer, C. (2000) 'The Akha of the Southwest China borderlands', in E. Spongel (ed.) *Endangered Peoples of Southeast and East Asia*, Westport, CT: Greenwood Press, pp. 37–53.

Kleinman, A. (2000) 'The violences of everyday life: the forms and dynamics of social violence', in V. Das, A. Kleinman, M. Ramphele and P. Reynolds (eds) *Violence and Subjectivity*, Berkeley: University of California Press, pp. 226–41.

Kleinman, A., Das, V. and Lock, M. (1996) 'Introduction', *Daedulus,* 125, 1: xi–xx.

Lyttleton, C. (2004) 'Relative pleasures: drugs, development and enduring dependencies in the Golden Triangle', *Development and Change,* 35, 5: 909–35.

Lyttleton, C. and Cohen, P. (2003) 'Harm reduction and alternative development in the Golden Triangle', *Drug and Alcohol Review,* 22: 83–91.

McCaskill, D. (1997) 'From tribal peoples to ethnic minorities', in D. McCaskill and K. Kampe (eds) *Development or Domestication: Indigenous Peoples of Southeast Asia,* Chiang Mai: Silkworm Press, pp. 26–60.

Nagengast, C. (1994) 'Violence, terror and the crisis of the State', *Annual Review of Anthropology,* 23: 109–36.

Nonini, D. (2002) 'Introduction: transnational migrants, globalization processes, and regimes of power and knowledge', *Critical Asian Studies,* 34, 1: 3–17.

Parfitt, T. (2002) *The End of Development,* London: Pluto Press.

Romagny, L. and Davineau, S. (2003) 'Synthesis of reports on resettlement in Long District, Luang Namtha Province, Lao PDR', unpublished report, Vientiane: Accion Contra la Faim.

Scott, H. (2002) 'Was there a time before race?: capitalist modernity and the origins of racism', in C. Bartolovich and N. Lazarus (eds) *Marxism, Modernity and Postcolonial Studies,* Cambridge: Cambridge University Press, pp. 167–184.

Sugar, J., Kleinman, A. and Heggenhougen, K. (1991) 'Development's downside: social and psychological pathology in countries undergoing social change', *Health Transition Review,* 1, 2: 211–20.

Sulak Sivaraksa (2002) 'Economic aspects of social and environmental violence from a Buddhist perspective', *Buddhist-Christian Studies,* 22: 47–60.

Taussig, M. (1984) 'Culture of terror – space of death: Roger Casement's Putumayo report and the explanation of torture', *Comparative Studies of Society and History,* 26: 467–97.

The Economist (2001) 'A violent mix, land and religion', 7 April: 32–3.

The Economist (2003) 'A survey of capitalism and democracy', 28 June: 1–16.

Toyota, M. (1996) 'The effects of tourism development on an Akha community', in M. Parnwell (ed.) *Uneven Development in Thailand,* Aldershot: Ashgate Publishing, pp. 226–40.

Tsing, A. (2000) 'The global situation', *Cultural Anthropology,* 15: 327–60.

UNDP (2001) *Lao National Human Development Report 2001: Advancing Rural Development,* UNDP: Vientiane, Lao PDR.

UNDP (2003) *Thailand Human Development Report 2003,* Bangkok: UNDP.

4 Constructions of foreign labour migrants in a time of SARS

The case of Singapore

Theresa Wong and Brenda S.A. Yeoh

Introduction

'Migration' and 'health' are undeniably geographical, in terms not only of their trajectories and movements over physical space but also of the place-specificities of health and migration policies (Kearns and Moon 2002: 610). In the geographical literature exploring the intersection between migration and health, several important shifts in the themes of interest have been noted in recent years. The traditional preoccupation in medical geography, well illustrated by the considerable work which has developed around AIDS, was with mapping the spatial diffusion of the *virus* or *disease* as a result of migratory activities (Kearns 1996: 124). This interpretation of medical geography has given way recently to a broader concern with 'geographies of health and health care' in which 'concern is with matters of individual and community identity, and with determining what elements of place shape individuals' [migrants' and non-migrants'] health experiences and access to services' (Mohan 2000: 496). One important strand in the burgeoning literature on 'geographies of health and health care' emphasizes the socio-biological construction and corporeal experience of health, as well as how socially constructed notions of 'disease', and the 'vectors' of disease, feature in the provision of, and access to, health care, as well as featuring in public health strategies in combating disease and ill-health (Browne and Barrett 2001; Parr 2002). Within this frame, the migrant is of special interest not only due to the acceleration of migration across borders in a globalizing world, but also because migrant workers – whether in terms of their entry into host societies or in terms of their return migration – have often been identified as a 'high-risk' disease vector (Weerakoon 1997: 72) which is 'more vulnerable than local populations' (*Migration and Health Newsletter*, International Organization for Migration, March 2002). Indeed, throughout the history of health and infectious disease control, the migrant, especially one from a less developed country, has often been the subject 'specifically classified and excluded from the territory and the body politic' of the nation (Bashford 2002: 344) and associated with a heavy infectious disease burden, despite the lack of scientific evidence (Tavira, n.d.).

Attitudes towards migrants in both public and government discourse, and particularly as enmeshed in local politics, set the tone for the treatment of migrants as contributors of disease and ill-health, and these attitudes can set in motion discriminatory policies that widen the rift between local society and migrant worker populations. Migrants are often perceived to be a mobile, floating and often 'disposable' population, bodily reservoirs of disease and infection. Such attitudes towards migrants tend to prevail, especially in cases where arriving migrants or foreign workers are attributed a lower economic status *vis-à-vis* the local workforce, in which class power relations are often tilted in favour of the host community. As the other chapters in this volume (see e.g. Asis, Chapter 7, this volume) also demonstrate, such perceptions of lower status migrants are often founded on classed constructions of ethnicity and nationality.

In periods of health crises and disease outbreaks, negative public attitudes towards migrants are often writ large, culminating in heightened suspicion and increased surveillance over migrant populations. Although it is often public groundswell that fuels fear and discontent, the authorities, in managing the crises and outbreaks, also contribute to producing images of migrant populations and bodies as conduits of infection. In fact, the focus on migrant workers and the tightening of immigration control is often a conscious strategy on the part of governments as a means of disease control; as was evident in many countries during the 2003 Severe Acute Respiratory Syndrome (SARS) outbreak (Gushulak 2003). The consequences of the tendency to blame foreigners, especially less skilled, low-status foreign workers, are twofold. First, focusing primarily on migrants as a high-risk group without fully understanding the social epidemiology of the disease (as is the case with 'new' diseases such as SARS) may lead to blind spots in the medical gaze of policy-makers with respect to other groups outside the category. Second, a knee-jerk reaction blaming migrants during high-profile health crises deepens misconceptions about migrants, which is neither productive for the curtailment of the disease nor effective in terms of the management of in-migration flows. Hence, it is important to unpack the relations of power and prejudice undergirding public discourses which couple disease with migration. By training the analytical lens on state policies and public discourses around the issue of a pool of more than 700,000 labour migrants in Singapore during the outbreak of SARS, we hope to clarify some of these relations woven into the way the SARS crisis was constructed.

New geographies of public health

The emphasis on the theme of power surrounding the control of disease and public health is rooted in the movements of social science and geographic research into the arena of 'new' public health in the past two decades. The shift in focus from curative to preventive forms of health care has led to the emphasis on 'risk', with public health practice in Western societies emphasizing individual responsibility in the prevention of ill-health and disease (Brown and Duncan 2002). The 'risk' or predisposition to a disease is seen to be caused by the 'combination of abstract factors' (Castel 1991: 287), rather than a specific peril or threat manifested in a singular person or group of people. According to some (Brown and Duncan 2002; Armstrong 1995), these definitions of a new social/health science have led to a popularizing of 'surveillance medicine', which is not only concerned with people with the actual disease, but extends the medical gaze into the spaces of everyday life to bring everyone into its fold. This extension of the network of medical surveillance also looks beyond life-styles and behaviours signalling health risks, to include an emphasis on the 'social and physical environment within which the individual body is placed' (Brown and Duncan 2002: 364). It is within the framework of this new turn in public health policy, research, and the intersection of the two, that this case study is situated.

In this context, we turn to the SARS crisis as it unfolded in Singapore. A few salient points about the crisis and the way it was managed need to be noted. First, although SARS has been termed an epidemic, its fatality and infection statistics do not come close to that typical of AIDS, cholera and influenza outbreaks. The death-toll from SARS during the March to June 2003 outbreak was 774 people, compared to 23,800 deaths from cholera in one month during the 1994 epidemic among Rwandan refugee camps in the Democratic Republic of the Congo (World Health Organization 2003), and the 'Hong Kong' flu outbreak from 1968 to 1970 which killed 45,000 people globally (Centre for Disease Control 2004). Yet the measures used to counter the spread of the SARS 'epidemic' in affected countries are disproportionate in their reach and extremely complex. More people have been affected by the preventive measures of SARS than by the SARS virus itself, which makes it an interesting case study for understanding the critical geographies of risk and disease vulnerability. Second, 'surveillance medicine' presumes the extension of a medical gaze with the aim of 'bring[ing] everyone within the network of visibility' (Armstrong 1995: 395). As will be discussed below, the SARS case demonstrates that while authorities indeed aim to have everyone brought into visibility, the preventive/medical lenses in Singapore are specifically trained on certain groups of people, particularly foreign workers assumed to be more 'vulnerable' to the disease. The epidemiological gaze on these workers is, at times, blind to the spaces they occupy within the local population, and yet in other circumstances the policies are extremely cognizant of space. It is this variable and inconsistent treatment of foreign workers that provides insights into the basic assumptions held about migrant workers in Singapore, often a product of their perceived social dissimilarity from, and at the same time spatial proximity to, the local Singaporean population.

The landscape of SARS infections and prevention

What then makes SARS different from other disease outbreaks? First, SARS is a relatively new disease whose unprecedented nature elicited a huge public outcry among populations and authorities (Xiang and Wong 2003). Second, not much was known about the aetiology of the disease in the earlier days, resulting in the need for extensive control measures to cover every eventuality. Third, and most significant, the rapid spread of SARS from its suspected origin in Guangdong, China to far-flung parts of the globe was due to the flows and movements of people, facilitated by what Glosserman (2003: 1) calls 'an international mobility that most of us take for granted'. It is an international mobility spurred on by changing international and national socio-economic conditions in response to globalization processes. However, just as the spoils of globalization are unevenly distributed, so are the benefits of this international mobility. Unskilled migrant workers, sometimes described as the 'underbelly' of global cities (Chang and Ling, quoted in Stasiulis and Bakan 1997: 121), are relegated to the lower undervalorized circuit of globalization, subject to tight and selective controls at the transnational border, treated as part of the 'buffer' sectors of the workforce, and vulnerable to immediate removal and repatriation during economic downswings (Yeoh and Chang 2001: 1032). Two different notions of 'vulnerability' are hence evident in the prevailing discourses: the 'vulnerability' of foreign worker populations to disease – sometimes presumed to be caused by the lack of ability to control their bodily movements – which is translated into the increasing 'vulnerability' of the local/home population – the second notion of vulnerability. We will now consider the divergent trajectories of the disease and its control – the epidemiological landscape of SARS, its infectors and spread; and the landscape of blame.

The first reports of the existence of a SARS-like pneumonia appeared in November 2002 in Guangdong, China, but it was not until March 2003 that the virus began to cause widespread international concern. Resembling atypical pneumonia, it was the prevailing view that SARS was spread primarily by contact with water droplets expelled from an infected person who coughed or sneezed. In November 2002, months before a global alert was raised in March 2003, the outbreak in Guangdong had already infected 792 people and caused thirty-one deaths. However, it was only when the disease crossed national borders that people began to take notice of what was thought of as a localized disease. The outbreak in Vietnam was the first to garner world attention when a man died in a Hanoi hospital, his condition worsening rapidly after being admitted for a high fever, dry cough, muscle ache and a sore throat from which he had suffered for four days. Following this, seven health care workers who had come into contact with the man fell ill with the same symptoms. In Hong Kong, the outbreak began on 11 March 2003 when a group of people in the Prince of Wales Hospital, including twenty-three health care workers, were found to have SARS. The situation worsened dramatically as the number of cases and deaths escalated, the fear culminating in reactions to the sudden news that 213 people, all from a single large housing estate (Amoy Gardens) with 15,000 residents, had suspected SARS. The outbreak in Singapore began with three Singaporean tourists returning from a trip to Hong Kong. By 7 April 2003, 106 people had SARS, and six people had died in the city-state. Fears surrounding these infections were further compounded by news of a number of cases beyond Asia, in Germany, Switzerland and Canada.

The idea that the virulent disease with a 20 per cent death rate could have spread so quickly from East Asia to Europe and North America gave rise to panicked reactions by governments. There seemed no question that it had to be handled top-down, from the highest levels of government. On 15 March, German customs officials quarantined a plane-load of 155 Singapore-bound passengers transiting in Frankfurt after a passenger was suspected to have been taken ill with SARS. As aspects of daily life began to be dominated by fear, and with the imposition of control measures such as the quarantine, certain countries began to emerge as 'hot zones' of SARS. While Europe's encounter with SARS turned out to be short-lived, with the number of cases remaining low and the threat of infections disappearing after a short while, the SARS outbreak was cited as 'the first real epidemic of a globalized Asia' – the battle had just begun for the Singapore, Hong Kong, Canadian and Chinese governments. Hong Kong's situation was severe – 1,755 cases and 282 deaths. Singapore was one of the countries worst hit by SARS, apart from China and Hong Kong, with 206 cases and thirty-three deaths over the three-month period until the outbreak was considered to have been contained.[1] Canada, the only country outside Asia which experienced a sustained rise in the number of cases, reported thirty-eight deaths and 250 cases by the end of the outbreak in July. By early August, the World Health Organization reported a total of 8,422 cases spread over thirty-two countries (Table 4.1).

Information about SARS, its causes and containment in the first month of the outbreak was scarce, leaving states to set guidelines for its containment. Some of the first targets of surveillance were migrant groups who were thought to have 'imported' the disease. The surveillance of migrant populations points to a more deeply embedded complex of attitudes within society and government over the question of migration, and in Singapore in particular, the dilemma over foreign workers.

Migrant worker policies in Singapore

Foreign workers in Singapore have been at the forefront of controversy in the past few years. Although the country has always depended on foreign workers – many regard the island-state as having been built by the toil of immigrants – the government has in the past decade espoused a particular openness to foreign workers. This welcoming attitude stems from a host of

Table 4.1 SARS cases by country, 1 November 2002 to 7 August 2003

Areas	Cumulative number of cases	Number of imported cases (per cent)
Australia	6	6 (100)
Brazil	1	1 (100)
Canada	251	5 (2)
China	5,327	NA
China, Hong Kong special administrative region	1,755	NA
China, Macao special administrative region	1	1 (100)
China, Taiwan	665	50 (8)
Colombia	1	1 (100)
Finland	1	1 (100)
France	7	7 (100)
Germany	9	9 (100)
India	3	3 (100)
Indonesia	2	2 (100)
Italy	4	4 (100)
Kuwait	1	1 (100)
Malaysia	5	5 (100)
Mongolia	9	8 (89)
New Zealand	1	1 (100)
Philippines	14	7 (50)
Republic of Ireland	1	1 (100)
Republic of Korea	3	3 (100)
Romania	1	1 (100)
Russian Federation	1	NA
Singapore	238	8 (3)
South Africa	1	1 (100)
Spain	1	1 (100)
Sweden	3	3 (100)
Switzerland	1	1 (100)
Thailand	9	9 (100)
United Kingdom	4	4 (100)
United States	33	31 (94)
Viet Nam	63	1 (2)
Total	8,422	–

Source: World Health Organization (2003).

factors, including government-led strategies for economic restructuring and a desire for Singapore companies to compete in the global market, an ambition that could not be fulfilled by the size and skills base of the local labour force alone. To date, approximately 18.8 per cent (or one in five persons, 754,524 people) out of the total resident population of four million people are foreigners, a statistic with immense repercussions for the economic and social landscape of the country. Public discourse, as exemplified in the Singapore newspaper dailies, is not short of expressions of discomfort at the sight of congregations of foreign workers on their off-days, and anxieties over whether local workers are losing high-end jobs to expatriate workers (Yeoh and Huang 1999).

With fertility rates on a probably irreversible downward trend, an ageing population, a small labour pool as well as brain drain, the local labour supply is expected to enter a long-term crisis situation and is unlikely to be able to sustain the government's economic ambitions for Singapore to become a hub for high-end financial and technology services (see Hui 1998). Attracting a foreign labour supply – both in terms of high-end 'foreign talent' and less skilled workers performing '3D' (dirty, dangerous and demeaning) work – is hence a necessary accompaniment to the city-state's globalizing ambitions. The majority of foreign workers in Singapore comprise unskilled or low-skilled workers in construction, service, manufacturing and the domestic work sectors, while a smaller but significant proportion belong to the skilled category of professionals in the financial and commercial sectors. In 2000, there were 470,000 low-skilled foreigners, compared to approximately 80,000 skilled workers and professionals. About 200,000 low-skilled foreign workers occupy jobs in the construction industry (mainly Malaysian, Thai and Bangladeshi), while another 140,000 are employed as domestic workers (predominantly Filipino, Indonesian and Sri Lankan). The skilled workers category has traditionally comprised foreigners from the USA, Britain, France and Australia, but skilled Asian expatriates, particularly from China and India, are fast becoming the dominant group. The foreign worker composition thus reflects the two-pronged approach of the government in its quest to solve Singapore's local labour problem, in conjunction with national ambitions to become an important player in the global economy. At the high end, workers who are globally marketable are paid wages that are competitively matched to salaries in other world cities, while at the low end, low-skilled workers are given wages with the regional labour market rates in mind (Chua 2003).

In official statistics and government management schemes, foreign workers are divided into categories primarily in accordance with the salary they earn.[2] Employment passes given to 'skilled' migrants are further subdivided into P passes and Q passes (Table 4.2). The top category of employment pass holders – P1 – draws a monthly salary of more than SGD$7,000[3] (GBP£2,194) to be able to 'qualify' for the employment pass; foreigners who are given such passes are generally advanced degree holders and have secured professional, administrative, executive or managerial jobs, or who are entrepreneurs or investors. Q passes, on the other hand, are generally issued to those with 'acceptable degrees, professional qualifications and specialist skills' (Ministry of Manpower 2002), and earning at least S$2,500 (GBP£784); in addition, to qualify, foreigners need to produce evidence of 'acceptable' degrees, professional qualifications or specialist skills (*The Straits*

Table 4.2 Different eligibility schemes for Employment Pass holders

Type of pass	Pass	Eligibility	Eligible for dependant's pass?[a]	Eligible for long-term social visit pass?[b]
P[b]	P1	For foreigners whose basic monthly salary is more than S$7,000.	Yes	Yes
	P2	For foreigners whose basic monthly salary is more than S$3,500 and up to S$7,000.	Yes	Yes
Q[c]	Q1	For foreigners whose basic monthly salary is more than S$2,500 and who possess acceptable degrees, professional qualifications or specialist skills.	Yes	No
	Q2	A Q2 pass is issued on *exceptional grounds* to foreigners who do not satisfy any of the above criteria. Such applications will be considered on the merits of each case.	No	No
R[d]	R1	For foreigners with National Technical Certificate (NTC)-3 practical certificates or suitable qualifications.	No	No
	R2	For unskilled foreign workers.	No	No

Source: Ministry of Manpower (2002), adapted by Yeoh and Wong (2003).

Notes
a Dependant's passes are issued to the children (under 21 years of age) and spouses of employment pass holders, entitling them to come to live in Singapore with the employment pass holder.
b P passes are for foreigners who hold acceptable degrees, professional qualifications or specialist skills, and who are seeking professional, administrative, executive or managerial jobs, or who are entrepreneurs or investors.
c The long-term social visit pass accords long-term visit entitlements to parents, parents-in-law, step-children, spouse (common law), handicapped children, and unmarried daughters over the age of 21.
d R pass holders are subject to a security bond and medical examination required for current two-year work permit holders.

Times, 9 November 2001). Holders of P and Q passes may work in any sector of the economy, are not subject to levies, and may bring family members with them to live in Singapore.

Unskilled or semi-skilled workers who do not fall into the above category in terms of salary earned are categorized as work permit holders (or R pass holders). They have few privileges and face greater restrictions, such as non-eligibility for the dependant's pass which would allow them to bring their spouses and children with them. In addition, they may not marry

Singaporeans, and are subject to medical examinations and employers' putting up a security bond. Table 4.2 details the categories of foreign workers and the varying levels of eligibility for the twin privileges – the dependant's pass and the long-term social visit pass (for family members) that would allow foreign workers to bring their families to Singapore during their stay.

Open borders discourse

The perception of foreign workers as a 'high-risk group' vulnerable to being infected with SARS, however, did not mute the discourse on the necessity of foreign workers in Singapore. The conflation of the SARS crisis and the discontent over the replacement of Singaporean workers with foreign workers became evident, with the worsening economy already reeling from the 1997 Asian financial crisis. SARS was estimated to have created losses of S\$1.5 billion (GBP£470 million) (Prime Minister Goh in *The Straits Times*, 1 May 2003), with GDP growth rates expected to fall by 2.3 per cent (Asian Development Bank estimates: see Fan 2003), losses that were expressed in some large-scale job retrenchments. While the SARS crisis ignited fresh attempts to discredit the foreign talent schemes, appeals by Singaporeans in the local newspapers seemed to point to a concern over the worsening economic recession made visible by a spate of retrenchments, rather than a concern over foreign workers' bodies as transmitters of infection. Although both Singaporeans and foreign workers were asked to leave, the rumblings of discontent over the persistent hiring of foreigners over locals were still audible.

However, whether companies were worried about their foreign workers being potential infectors, or whether Singaporeans were disgruntled about the continued hiring of foreign workers even as locals were being laid off, the government remained adamant that the country's borders could not be closed – that even SARS could not be a justification for keeping foreign workers away given their importance to the national economy. Ministers stoutly defended the policy, urging Singaporeans to understand that the long-term survival of Singapore depended on its continued engagement with foreign labour and foreign skills. At the height of the SARS crisis in April, for example, the Minister for Manpower, stressing the need for companies to retain their workers and not to send them away because of SARS fears, argued that 'if you take the approach of building a wall around Singapore, at the end of the day, you might find that everything comes to a standstill' (*The Straits Times*, 11 April 2003).

'Fighting SARS together'?

Singapore's first contact with SARS came in the form of two Singaporean women who returned from Hong Kong on holiday with the virus. The victims were hospitalized, and although they did not succumb to the disease, the virus was passed on to friends and family members with whom they came into contact, as well as to some hospital staff. It was deduced that in order to curb the spread of the virus, infected persons had to be isolated from other people, and that hospital staff must wear protective gear when treating SARS patients. The Singapore government responded quickly to the new information. On 24 March 2003, the *Infectious Diseases Act* was amended to give authorities greater powers to quarantine and to isolate all individuals having close contact or suspected of having SARS. New directives went out to companies, educational institutions and service sector employers, intensifying surveillance over the population, both in the public and private arenas. In the private (or quasi-private) spaces of homes and workplaces, individuals were encouraged to monitor their own health and to take measures according to their evaluation of their symptoms (for example, a fever of above 38 degrees Celsius, a dry cough and breathing difficulties). People reporting for work in public buildings were met with either temperature checks at the front door or directives to take their own temperatures at least twice a day. In what was a comprehensive and costly exercise, thermometers were issued to every home and every employee. In addition, individuals were discouraged from going to places where large crowds of people tended to congregate. Finally, in a typically Singaporean fashion, a public campaign – 'Fighting SARS together' – was launched to spur the population on to practise good hygiene, both at home and in public, as an individual and collective responsibility.

While the local population was urged to unite in the fight against SARS, the large foreign worker community was subject to a hierarchical system whereby SARS containment and prevention measures were imposed to different degrees according to the type of employment permit held by the migrant. This system was a finely detailed, complex delineation of differential measures, achievable only in a context like Singapore where tight immigration controls were already firmly in place prior to the SARS onslaught. The rules for foreign workers (Table 4.3) were laid down on the basis of the designated observation period (otherwise known as the quarantine period) of ten days. This was the number of days recognized as the maximum amount of time needed for SARS symptoms to show up after an infection took place.

On the one hand, work permit holders (R pass holders), with the exception of foreign domestic workers, were subject to quarantine at facilities built especially for the purpose. The facilities comprised dormitory beds, located in the far-flung parts of the island (industrial areas) away from the general population. On the other hand, employment pass holders returning from SARS-hit areas had to observe a similar ten-day observation period 'at the company or other isolation facilities pre-arranged by [the] employer'. This usually meant the private residence of the employment pass holder. However, what was perhaps most indicative of the assumptions held about work permit and employment pass holders were the measures for existing foreign workers returning to Singapore from SARS-hit areas. While return-

Table 4.3 Observation guidelines for Work Permit (R pass) and Employment Pass (P and Q pass) holders

Foreign worker	Need for observation period	Observation period	Notes
Work permit (WP) holders			
New WP holders arriving from SARS-hit areas	Yes	10-day observation period at JTC designated isolation quarters	After the 10-day period, foreign workers will be released and subject to medical clearance. Employers should then follow normal procedures to obtain the work permits
Existing WP holders leaving for SARS-hit areas	Yes, upon return to Singapore	10-day observation period at JTC designated isolation quarters	WP will be cancelled once the WP holder leaves for SARS-hit area. Employers will have to make fresh WP applications for these workers and they will be subject to the 10-day observation period on entering Singapore
Employment pass (EP) holders 'P' pass holders			
New 'P' pass holders and related pass holders arriving from SARS-hit areas	Yes	10-day observation period at company or other isolation facilities prearranged by employer	Passes will be issued only after compliance with all observation period procedures
Existing 'P' pass holders and related pass holders returning from SARS-hit areas	No	NA	NA
'Q' Pass Holders			
New 'Q' pass holders and related pass holders as well as training visit pass holders arriving from SARS-hit areas	Yes	10-day observation period at company or other isolation facilities prearranged by employer	'Q' and related passes will be issued after compliance with all observation period procedures
Existing 'Q' pass holders and related pass holders as well as training visit pass holders leaving for SARS-hit areas	Yes, upon return to Singapore	10-day observation period at company or other isolation facilities prearranged by employer	'Q' and related pass holders who do not comply with the observation period measures upon return to Singapore will have their passes cancelled

Source: Ministry of Manpower (2003a).

ing P pass holders did not need to undergo an observation period (even if they had gone to a SARS-hit country), R pass workers not only had to undergo the ten-day observation period, but had their work permits automatically cancelled when they left, even temporarily, for a SARS-hit country, and had to reapply for new permits upon their return. The official reason given for the exemption for P pass holders was that these workers were 'deemed to possess greater control over their travel requirements and thus able to avoid travel to SARS affected areas' (Ministry of Manpower 2003a). They were also deemed to be able to 'take their own precautions' (*The Straits Times*, 11 April 2003). Not only did this presume that P pass holders were more capable of exercising 'social responsibility' (Ministry of Manpower 2003a) than were R pass workers; differential surveillance further entrenched pre-existing disparities in rights accorded to migrant workers of different skill levels (for example, employment pass holders could bring their spouses as dependants to Singapore while work permit holders were not afforded this privilege).

P and R pass holders were evidently positioned at the two extreme ends of the spectrum of foreign workers, defined by the types of skills and the level of income earned. Q pass holders occupied the middle ground, and accordingly the mode of surveillance was calibrated to reflect this 'in-between' position: on the one hand, unlike the case with work permit holders, existing Q pass holders did not have their passes automatically cancelled on leaving for a SARS-hit country; on the other hand, they had to observe quarantine regulations on return to Singapore, unlike their P pass counterparts. The finely calibrated variations in the rules of surveillance clearly reflected differential assumptions about the degree of 'risk' posed by migrant bodies of different skill levels.

Spaces of work too had implications for the construction of the level of risk posed by the foreign worker in the campaign against SARS. While construction workers were often domiciled in communes located physically separate from the local Singapore population, another category of R pass workers – the domestic worker – was sequestered in the private household. By virtue of the space they occupied, domestic workers, who form a significant part of the foreign worker statistics, underwent a different kind of surveillance from other low-skilled or unskilled foreign workers, in spite of belonging to the same R pass category. Foreign domestic workers fell within the unit of the household, and if they were suspected of SARS or issued with a ten-day quarantine order, such quarantine periods had to be observed at the home of their employer. The government authority on this matter – the Ministry of Manpower – also stipulated that domestic workers should not be sent away because of fears of SARS – possibly to avoid the unnecessary circulation and mobility of these workers during the outbreak.

The idea that foreign domestic workers and construction workers should be subject to different measures was based on their respective spaces of work which, in turn, have roots in the gendered policies that govern them in general. Huang and Yeoh (2003: 92) showed that both groups of workers are governed under different ideological bases, and their work valorized differently, resulting in differential access to legal protection in the practical sphere, and dissimilar levels of social control. While male construction workers' work is assumed to be productive labour, encompassing the use of 'skills' and leading to tangible results, domestic work is not perceived as 'real work'. Likewise, unlike construction workers who were domiciled separately as a body of people away from local residential areas, foreign domestic workers were treated as part of the family unit, to be quarantined at the employer's home if necessary (Ministry of Manpower 2003b). Hence it was the 'host' family that was expected to exert control over the foreign domestic worker and to ensure that preventive measures were observed by the worker.

It is not surprising, therefore, that this stipulation became a source of tension between local Singaporean employers and foreign domestic workers. Letters to the main English daily during this time reflected the anxieties of some employers whose main fears were that domestic workers could infect their families because of close proximity and contact with family members. One reader suggested in *The Straits Times* forum page that many maids take their wards to school, and their congregation in such public spaces 'could be a source of SARS infection' (*The Straits Times*, 16 April 2003). Although these opinions were criticized by other Singaporean readers for their discriminatory undertones, they reflected some of the deep-seated prejudices and classed/ethnicized constructions of irresponsibility held towards foreign workers, which may also be found in public complaints about the spaces that foreign workers occupy – both physical and social – especially visible in the large sizes of their congregations on Sundays (their day off), in the 'weekend enclaves' of Little India, Orchard Road and Golden Mile.

Ironically, domestic workers might have had more reason to fear infection than their employers, as one maid was suspected to have caught the SARS virus from her infected employer (*The Straits Times*, 11 April 2003) – she later died. In Hong Kong, there were rumblings of discontent among domestic workers who felt that their employers had used the SARS crisis to deny them rights such as days off, but at the same time forced them to continue activities such as taking their children to school and going to the market, which could equally well have exposed them to the risk of catching the virus (*Agence-France Presse*, 7 May 2003). Foreign domestic workers, by virtue of their proximate relations with their employer-families, are perhaps the most vulnerable migrant worker group of all. As of 1 May 2003, at the height of the disease, eleven Filipino and seventeen Indonesian domestic workers had been infected in Hong Kong (with one death later) and one Filipino domestic worker had died in Singapore. This is in contrast to the number of non-foreign domestic-worker victims; one SARS death in Canada of a Filipino caregiver in a semi-domestic capacity, and one infected Indian migrant worker in Singapore (Coalition for Migrants' Rights 2003).

As Gushulak (2003) has pointed out, foreign workers are often the first to be targeted in control measures during disease outbreaks. Reasons given by Singapore authorities about the relatively strict controls on foreign workers included their mobility, their tendency to congregate, and their assumed relative lack of social responsibility towards the country of their temporary domicile. According to the Ministry of Manpower (2003a), authorities applied 'a higher level of checks' to foreign workers than to tourists to SARS-hit countries 'because [foreign workers] will live and work here and mingle with the Singaporean community on a long-term basis'. This line of reasoning is questionable, since Singaporean tourists abroad, who are not subject to the same draconian measures, also return home to their families and friends, with a greater potential to infect other Singaporeans. Although discouraged from doing so, Singaporeans were still free to visit SARS-hit countries for tourism, as well as for business. China and Hong Kong are two of the most important places for Singaporean overseas travel. Moreover, tourist flows (both incoming and outgoing) were not tracked with the same degree of surveillance as were the movements of foreign workers. On 12 June 2003, the Singapore Tourism Board, a government agency responsible for the promotion of Singapore as a tourist destination, announced a S$200 million programme to boost tourist numbers which had been severely hit by the SARS crisis. One of the most sought-after groups of tourists were those from China. Arrivals from China had been steadily increasing, with a 14.5 per cent rise in 2001 alone, making Chinese visitors the brightest spot in terms of tourist arrivals in recent years (Singapore Tourism Board 2001). However, as of the date of the announcement of the tourism strategy, China had not obtained the all-clear signal from the WHO.[4] Ironically, China had imposed a travel ban to Singapore in mid-April 2003 when the city-state was in the throes of the SARS crisis, causing considerable dismay in Singapore over the plummeting Chinese tourist numbers, which failed to revive even with the lifting of the ban a few weeks later (on 27 May for Beijing residents, and on 9 July for the rest of China). Immediate efforts were put into wooing the Chinese back to Singapore, including the waiving of the security deposit for individual tourists from China, and when the first Chinese tourist group set foot in the city-state, they were 'warmly welcomed ... by celebrations of lion dance, orchid garlands, and goodie bags' (*Singapore Travel News*, 31 July 2003).

There is also no compelling reason why tourists, whether Singaporean or not, should possess a greater degree of social and moral responsibility towards others, and towards curbing the spread of the virus in Singapore. Indeed, foreign workers may have a greater incentive to keep themselves SARS-free for fear of losing their economic livelihoods. When a special break of ten days was declared for all schoolchildren in Singapore during the upsurge of SARS cases and deaths, authorities found their attempts to detect, identify and isolate new cases of SARS thwarted by parents taking advantage of the break to take their children on overseas vacations, even to SARS-hit countries such as China and Hong Kong, instead of being mindful of the principle of a ten-day recess from school – which was to allow for SARS symptoms among schoolchildren to show up safely outside the crowded school environment. The more stringent control of migrant, rather than tourist, bodies was somewhat ironic, given that Singaporean tourists in Hong Kong had first brought the virus into Singapore – the first SARS victims being three Singaporean women who had returned from a holiday in Hong Kong.

Conclusion

Migrant populations, many already subject to repatriation during economic downturns, are often the first targets of epidemiological surveillance in disease outbreaks. Even while the relationship between migration and health remains extremely tenuous, health security concerns become a ready platform for expressing deep-rooted prejudices against migrant workers in public discourse, as well as a justification for strengthening the role of the state as gatekeeper of the nation through its finely governed, selective control of transnational borders. It is during a crisis such as SARS that migrant workers are exposed as mobile, transient and not always controllable, requiring the implementation of quarantine measures and other solutions to shut out the killer virus and prevent it from crossing national borders.

The SARS crisis in Singapore revealed a host of complex and finely differentiated policies based on particular assumptions about migrant bodies. As was discussed above, surveillance strategies employed to oversee foreign domestic workers (all female) and foreign construction workers (all male) also differed on the basis of their spaces of work, which in turn are governed by gendered assumptions about the 'place' of men and women. Furthermore, in a time of SARS, surveillance strategies discriminate between migrants and nationals, reflecting the underlying higher predisposition of migrant groups towards the disease. This higher level of risk is presumed to originate from a set of conditions, including the physical spaces and socio-economic positions that migrant workers occupy, their assumed behaviours, and the pre-existing categorization between classes of foreign workers. Lines of restriction drawn for migrant worker categories did not exist in the same way for the local Singaporean population, whose freedom of movement, especially for tourism, was only subject to notices of caution to exercise self-responsibility. A second set of boundaries are defined between 'classes' of foreign workers. Differential restrictions within the foreign worker population give high-end labour migrants the benefit of the doubt by allowing them the freedom to control their own bodies and movements, while imposing high penalties on low-end migrants, whose movements are already restricted by the kind of permit they hold when they first arrive.

The transnational border is therefore a finely controlled mechanism policed by the gatekeeper, the nation-state, which insists at the same time that the country's borders cannot be closed – that even SARS cannot be a justification for keeping foreign workers away because of their importance to the national economy. The SARS crisis brings out the rigid immutability of the boundaries between classes of foreign workers, and at the same time highlights the sieve-like nature of the national border. The different configurations of regulations imposed on groups of workers show that policing the border is a complicated and difficult task, but one that nation states seize upon as crucial in proclaiming war on SARS, disease and diseased bodies.

Notes

1 The WHO maintains a list of countries which remain on alert for the appearance of new SARS cases. Singapore was deemed a risk region until 2 June 2003, when no new infections were detected.
2 Foreign worker statistics cited in the text refer only to those who are legally employed. While the 'immigrant offender problem' – comprising both illegal immigrants and those who 'overstay' their work permits – is a serious one in Singapore, it appears that the problem is kept under control by the authorities. The onslaught of SARS further stiffened vigilance against illegal entry. The number of illegal immigrants (mainly from India, Myanmar and China), for example, fell by 30 per cent from 7,860 illegal immigrants in 2002, to 5,510 in 2003, possibly as a result of the deterrent effects of the increased severity of penalties for illegal entry (including a gaol term and caning), as well as stepped-up enforcement actions and security checks at checkpoints, the latter significantly amplified in the aftermath of the 9/11 incident (Immigration & Checkpoints Authority, http://app/ica.gov.sg/pressrelease_view.asp?pr_id=202, accessed 8 July 2004). As a specific SARS control measure, The Immigration and Checkpoints Authority and the Ministry of Manpower issued a joint statement (dated 26 May 2003) urging illegal immigrants and overstayers with possible SARS symptoms to come forward voluntarily to seek treatment, on the assurance that their immigration or illegal employment offence would be compassionately considered. Foreign workers were also urged to report any of their friends with symptoms of SARS to the authorities (Immigration & Checkpoints Authority, http://app.ica.gov.sg/pressrelease/pressrelease_view.asp?pr_id=149, accessed 8 July 2004).
3 One Singapore dollar (SGD$1) was about GBP£0.31 at the time of writing (17 February 2004).
4 China, which had the highest death-toll from SARS in the world, was taken off the WHO list of areas with local SARS transmission in early July 2003.

References

Agence-France Presse (2003) in INQ7.net. online, available at: www.inq7.net/brk/2003/may/07/brkofw_1–1.htm (accessed 2 September 2004).

Armstrong, D. (1995) 'The rise of surveillant medicine', *Sociology of Health and Illness*, 17: 393–404.

Bashford, A. (2002) 'At the border: contagion, immigration, nation', *Australian Historical Studies*, 120: 344–58.

Brown, T. and Duncan, C. (2002) 'Placing geographies of public health', *Area*, 33, 4: 361–9.

Browne, A. and Barrett, H. (2001) 'Moral boundaries: the geography of health education in the context of the HIV/AIDS pandemic in Southern Africa', *Geography*, 86, 1: 23–36.

Castel, R. (1991) 'From dangerousness to risk', in G. Burchell, C. Gordon and P. Miller (eds) *The Foucault Effect*, London: Harvester Wheatsheaf, pp. 281–98.

Centre for Disease Control (2004) Basic information about Avian Influenza – background on pandemics. Online, available at: www.cdc.gov/flu/avian/facts.htm (accessed 6 January 2004).

Chua, B-H. (2003) 'Multiculturalism in Singapore: an instrument of social control', *Race and Class*, 44, 3: 58–77.

Coalition for Migrants' Rights (2003) *CMR-AMC May Day Statement on SARS*, cited in Asian Migrants online, available at: www.asian-migrants.org/resources/105221595610632.php (accessed 16 February 2004).

Fan, E. (2003) SARS: *Economic Impacts and Implications*, ERD Policy Brief No. 15, Manila: Asian Development Bank, Economic Research Department.

Glosserman, B. (2003) 'Why Asia should unite against SARS', Asia Report 14, 17 April 2003, Japanese Institute of Global Communications online, available at: www.glocom.org/special_topics/asia_rep/20030417_asia_s14/ (accessed 5 November 2003).

Gushulak, B. (2003) 'Population mobility, health and SARS: lessons from the past, lessons from today', *Metropolis World Bulletin*, 3: 13–15.

Huang, S. and Yeoh, B.S.A. (2003) 'The difference gender makes: state policy and contract migrant workers in Singapore', *Asian and Pacific Migration Journal*, 12, 1–2: 75–97.

Hui, W-T. (1998) 'The regional economic crisis and Singapore: implications for labour migration', *Asian and Pacific Migration Journal*, 7, 2–3: 187–218.

Kearns, R. (1996) 'AIDS and medical geography: embracing the other?', *Progress in Human Geography*, 20, 1: 123–31.

Kearns, R. and Moon, G. (2002) 'From medical to health geography: novelty, place and theory after a decade of change', *Progress in Human Geography*, 26, 5: 605–25.

Migration and Health Newsletter (2002) *International Organization for Migration*, March 2002

Ministry of Manpower, Singapore (2002) *Ministry of Manpower, Singapore*. Online, available at: www.mom.gov.sg (accessed 13 November 2003).

Ministry of Manpower, Singapore (2003a) *FAQs: Observation Period: General*. Online, available at: www.mom.gov.sg/MOM/CDA (accessed 16 November 2003).

Ministry of Manpower, Singapore (2003b) *Health Guidelines on Severe Acute Respiratory Syndrome (SARS) for Employers of Foreign Domestic Workers*. Online, available at: www.mom.gov.sg/MOM/CCD/Others/Guidelines_FDWs(Employers)_30Apr.pdf (accessed 6 September 2004).

Mohan, J. (2000) 'Medical geography', in R.J. Johnston, D. Gregory, G. Pratt and M. Watts (eds) *The Dictionary of Human Geography*, Oxford: Blackwell, pp. 494–6.

Parr, H. (2002) 'Medical geography: diagnosing the body in medical and health geography, 1999–2000', *Progress in Human Geography*, 26, 2: 240–51.

Singapore Tourism Board (2001) *Annual Report on Tourism Statistics 2001*, Singapore: Singapore Tourism Board.

Singapore Travel News (2003) 31 July. Online, available at: www.travel-singapore.com/news/Jul1.htm (accessed 16 July 2004).

Stasiulis, D. and Bakan, A.B. (1997) 'Negotiating citizenship: the case of foreign domestic workers in Canada', *Feminist Review*, 57: 112–39.

Tavira, L.T. (n.d.), 'Transmittable diseases of migrants: myths, facts and Epi-Migra', The European Project AIDS & Mobility, Netherlands Institute for Health Promotion and Disease Prevention, online, available at: www.aidsmobility.org/luis.doc (accessed 15 July 2004).

The Straits Times, various issues.

The Sunday Times, various issues.

Weerakoon, N. (1997) 'International female labour migration: implications of the HIV/AIDS epidemic in the Asian region', in G. Linge and D. Porter (eds) *No Place for Borders: the HIV/AIDS Epidemic and Development in Asia and the Pacific*, St Leonards: Allen & Unwin, pp. 67–77.

World Health Organization (2003) 'Cholera and epidemic-prone diarrhoeal diseases'. Online, available at: www.who.int/csr/disease/cholera/en/ (accessed 7 January 2004).

Xiang, B. and Wong, T. (2003) 'SARS: public health and social science perspectives', *Economic and Political Weekly*, 38, 25: 2480–3.

Yeoh, B.S.A. and Chang, T.C. (2001) 'Globalising Singapore: debating transnational flows in the city', *Urban Studies,* 38, 7: 1025–44.

Yeoh, B.S.A. and Huang, S. (1999) 'Spaces at the margins: migrant domestic workers and the development of civil society in Singapore', *Environment and Planning A*, 31: 1149–67.

Yeoh, B.S.A. and Wong, T. (2003) 'Foreigners and migration policy in Singapore', in W. Gieler (ed.) *Handbuch der Auslander – und Zuwanderungspolitik – Von Afghanistan bis Zypern (Handbook of Foreigners and Migration Policy)*, New York: Lit-Verlag, pp. 567–75.

5 Investigating the role of nativity on functional disability among older adults in Singapore

Santosh Jatrana and Angelique Chan

Introduction

Although there has been an increased focus on the relationship between migration and health, there is still limited knowledge of the effect of nativity on the health status of older adults. In spite of the possibility that there may be greater health problems and fewer resources among older foreign-born individuals than among the native-born, research has generally not incorporated nativity as a factor contributing to health status. Most studies on nativity deal with the health of young foreign-born people; studies of the health status of foreign-born older adults are scarce. Understanding nativity differences in health at older ages is particularly important in societies where a large proportion of the older population is foreign-born, as in the case of Singapore. At present, foreign-born migrants comprise approximately 40 per cent of the older adult population (fifty-nine and above) in Singapore (see Table 5.1).

The health status of older adults is of particular interest because the elderly usually command medical resources out of proportion to their number. According to Mayhew (1999), the number of people with disabilities is likely to grow substantially as societies age, and consequently ageing is poised to overtake population growth as the main factor for expanding health expenditures on a worldwide basis. Recent gerontological studies conducted in the USA indicate that health and social care services required by older people with disabilities will be a growing burden and a major societal concern for the next century (Fried and Guralnik 1997). The costs associated with health problems at older ages will be higher in Asian societies such as Singapore and Japan where the proportion of those in older ages is greater. To our knowledge, this is the first Singaporean study which focuses on differences in functional disability among older adults born in Singapore, and those born in China, Malaysia and elsewhere.

The aim of this chapter is to analyse the influence of country of birth on self-reported functional disability. The question we explore is whether nativity is an independent risk factor for reporting a functional disability, or whether nativity differences in self-reported functional disability merely reflect differences in socio-economic and demographic characteristics. This is

Table 5.1 Descriptive statistics by nativity status, 1999 (weighted percentages)

Variable	Native-born		China-born		Malaysia-born		Others	
	N	Column percentage	N	Column percentage	N	Column percentage	N	Column percentage
Reporting any disability								
No	680	65.3	288	46.9	153	64.6	38	46.3
Yes	361	34.7	326	53.1	84	35.4	44	53.7
Age								
Adults aged 55–69	576	55.4	66	10.7	148	62.4	31	37.8
Adults aged 70–79	296	28.5	155	25.2	58	24.5	25	30.5
Adults 80 and over	168	16.2	393	64.0	31	13.1	26	31.7
Sex								
Male	431	41.4	234	38.1	111	46.8	48	57.8
Female	610	58.6	380	61.9	126	53.2	35	42.2
Ethnicity								
Chinese	899	87.1	614	100.0	152	65.8	27	33.8
Indian	28	2.7	0	0.0	9	3.9	29	36.3
Malay	105	10.2	0	0.0	70	30.3	24	30.0
Marital status								
Currently married	524	50.4	219	35.6	144	60.8	51	62.2
Widowed/divorced/separated	476	45.8	386	62.8	80	33.8	31	37.8
Never married	40	3.8	10	1.6	13	5.5	0	0.0
Highest education level achieved								
Completed secondary and above	114	11.0	28	4.6	44	18.6	16	19.5
Completed primary	276	26.5	104	16.9	56	23.6	25	30.5
No education	650	62.5	483	78.5	137	57.8	41	50.0

	n	%	n	%	n	%	n	%
Average monthly income of respondent and spouse								
More than S$1,000	287	27.6	82	13.3	80	33.6	20	24.4
Less than S$1,000 (including none)	754	72.4	533	86.7	158	66.4	62	75.6
Adequacy of income								
Yes	834	80.2	527	85.7	191	80.6	74	90.2
No	206	19.8	88	14.3	46	19.4	8	9.8
Smoking								
Never smoked	669	64.3	404	65.8	154	65.3	56	68.3
Quit smoking	191	18.3	124	20.2	50	21.2	18	22.0
Current smoker	181	17.4	86	14.0	32	13.6	8	9.8
Alcohol consumption								
Never drinks	803	77.1	484	78.8	198	83.5	64	78.0
Past drinker	93	8.9	67	10.9	10	4.2	3	3.7
Current drinker	145	13.9	63	10.3	29	12.2	15	18.3
Emotional support								
Quite good	447	43.0	238	38.8	132	55.5	54	65.9
Some	179	17.2	92	15.0	36	15.1	11	13.4
Very little/not at all	414	39.8	284	46.3	70	29.4	17	20.7
Organisation membership								
Yes	108	10.4	87	14.2	20	8.4	9	10.8
No	933	89.6	527	85.8	217	91.6	74	89.2
Living arrangements								
With spouse/children/in-laws/other persons	978	94.0	571	93.0	226	95.4	78	95.1
Alone	62	6.0	43	7.0	11	4.6	4	4.9
Poor self-assessed health								
No	572	54.9	274	44.6	122	51.5	28	34.1
Yes	469	45.1	340	55.4	115	48.5	54	65.9

Source: *Transitions in Health, Wealth, and Welfare of Elderly Singaporeans: 1995–1999* survey.

Note

Percentages are weighted to account for over-sampling of Indians and individuals aged 75+ in the original 1995 survey.

an important question for policy-makers. If nativity differences in self-reported functional disability can be attributed to demographic and socio-economic characteristics, rather than to nativity itself, then functional disability reduction programmes can be targeted to compensate for the differences between groups based on traits such as education, employment and income. If, however, nativity itself is a significant determinant of functional health, then a culturally based barrier may exist, in which case simply expanding the availability of health services or altering people's socio-economic conditions may not enhance functional health. The findings of the study are particularly timely, given the trends in economic globalization that have increased the growth of the foreign-born populations in many countries, including Singapore, during the past few decades.

Why study nativity?

In future, many older adults will grow old in countries in which they were not born, hence studying the impact of nativity on functional health at older ages is very necessary. Migration is often correlated with greater stress associated with assimilation into a new country and racial discrimination, and lower socio-economic status (Kuo and Tsai 1986). While migration may offer economic benefits, physical health problems and psychological distress often result from the social stress involved in moving from one's country (Chung and Kagawa-Singer 1993; Shuval 1993), reduced security in daily life (Sundquist 1994), and experiences of alienation and discrimination (Kaplan and Marks 1990). Recent years have witnessed increased interest, particularly in Europe and the USA, in the health of foreign-born people, due in large part to changing population composition. But most of the previous research has treated foreign-born as one category, thus masking the important variations in health within the foreign-born owing to differences in socio-economic or socio-demographic issues. In the USA, for example, being foreign-born has been found to be associated with lower income, educational and occupational status compared to native-born adults, and these characteristics have been found to be generally linked to poorer health and fewer medical resources (Keefe *et al.* 1979; Keefe 1982; Alston and Aguirre 1987). Subsequent analysis of foreign-born migrants in different ethnic groups has revealed that older Mexican Americans have a higher prevalence of certain chronic conditions (for example, diabetes and obesity) that have been associated with disability (Hazuda and Espino 1997; Ostir *et al.* 1998).

We go beyond the dichotomous division of nativity into foreign-born and native-born, and deal with country of birth as an independent factor for predicting functional disability among older heterogeneous adults in Singapore. Several studies in Sweden have demonstrated that non-European (Latin American) refugees have higher rates of long-term illness even after controlling for social factors (Sundquist 1995). Men and women of Greek and Yugoslav origin have higher use of sickness and disablement pension, and

Southern Europeans, Chileans, Iraqis and Syrians have higher rates of long-term certified sick leave (greater than thirty days) (Kindlund 1995; Sundquist and Johansson 1997a) compared to people born in Sweden. Approximately half of all long-term certified sick leave among immigrants in Sweden can be predicted using occupation, the rest by factors connected with their migration process, particularly for married immigrants from South Europe (Kindlund 1995). These differences can also be partly explained by differences in socio-economic status and working environment, since immigrants are over-represented among the unemployed and in manual professions (Axen and Lindstrom 2002). However, even after adjusting for life-style and material factors, Sundquist and Johansson (1997a) found higher rates of severe long-term illness for males and females born in Eastern Europe and non-European non-Western countries, compared to the Swedish-born. The Sundquist and Johansson (1997a) study also demonstrates that foreign-born people from non-Western non-European countries were highly educated, but lived in a marginal social and cultural sector in Swedish society.

A comprehensive immigrant mortality study of England and Wales from 1970 to 1978 showed that ethnicity, defined as being a foreign-born minority, influenced mortality independent of social class (Marmot *et al.* 1984). The study found that the standardized mortality ratio (SMR) from hypertension among West Indians, diabetes in Indians, infections in Irish and Indians, cirrhosis of the liver and accidents among the Irish, and complications of pregnancy and childbirth among women born in the Indian subcontinent, Africa and the Caribbean, were all larger than the SMR for the most disadvantaged native-born social class in England and Wales (Marmot *et al.* 1984). These results are supported by longitudinal work as well. Raftery *et al.* (1990) used longitudinal survey data to show the persistence of a poorer mortality experience among residents of England and Wales with one or both parents born in Ireland (controlling for social demographic characteristics), compared to residents with two native-born parents.

It is not only the physical health but also the mental health of ethnic minority foreign-born that is affected by migration-related factors in the new country or cultural context. Several studies in Sweden, for example, have demonstrated that immigrant groups have an increased risk of attempted suicide (Sundquist 1994; Johansson *et al.* 1997; Bayard-Burfield *et al.* 1998), suicide (Johansson *et al.* 1997), psychological distress, and psychosomatic complaints (Bayard-Burfield *et al.* 1999). It could be the conditions of marginality or alienation from mainstream ethnic groups that increase mental disorders, or it could be the conditions of cultural difference under which immigrants must operate (for example, not knowing the language or the culture of the mainstream ethnic group) that makes them particularly vulnerable to emotional stress.

There are differences between the native-born and foreign-born immigrant groups not only in health status, regardless of how it is measured, but also in the use of health care services, although the evidence is mixed. Some studies have demonstrated that immigrant groups consume fewer health services in general (Stephenson 1995), while other studies have shown that immigrant groups use as much as, or more, health care services as do the indigenous population (Bowling *et al.* 1992).

Despite all these studies, the relationship between nativity and health has been a contentious issue. Apart from the contradictory findings, nearly all previous research has been conducted in the Western industrialized world, particularly in the USA and Europe. Consequently, very little is yet known about the association between nativity and health in non-Western settings where cultural norms and values differ from those that are commonly found in Western societies. Moreover, most of the research focuses on younger adults, and research on the elderly in developing countries has been rare. Although several recent studies have dealt with health status in a few Asian and African countries (e.g. Liu *et al.* 1995; Lamb 1997; Zimmer *et al.* 1998, 2002), none of these studies has considered nativity as a factor. Whether nativity differences at older ages in health status exist in Singapore, and the size and determinants of these differences, remain urgent empirical questions given the fact that immigration has played a significant role in Singapore's history.

Prior to 1819, when Sir Stamford Raffles established Singapore as a British trading station, the population of Singapore may have been approximately 200, made up of *Orang Laut* (sea people), Muslim Malay fisher people, and Chinese pepper and gambier cultivators (Bloom (1986) in Erb 2003). Once Singapore was established as a free port, migrants, mainly from China and India, came to make their fortune in Singapore with the aim of quickly returning home. However, events in their home countries (for example, the communist takeover of China in 1949) meant that many of these immigrants decided to make Singapore their home (Chan and Tong 2003). After Singapore's independence in 1965 and its separation from Malaysia, Malaysians living in Singapore acquired immigrant status. Thus, Singapore's immigrant population is multi-ethnic and multi-cultural. As Singapore is a small state with few natural resources, human resources are critical and the Singapore government has recruited foreign talent to meet shortfalls in human resources. However, unlike their predecessors who worked mainly as labourers or traders, today's immigrants are largely professionals. Foreign talent currently makes up 9 per cent of the professional workforce and has high productivity. Between 1991 and 2000 foreign professionals contributed 37 per cent to Singapore's gross domestic product. Thus it is no surprise that foreign professionals are often granted permanent residence status in Singapore to maintain population levels and boost human resource productivity. At present, Singapore's total fertility rate is 1.26 (Department of Statistics 2004) and immigration may play an increasingly important role in managing Singapore society.

Methods and materials

The sample

We use cross-sectional data from the 1999 survey of *Transitions in Health, Wealth, and Welfare of Elderly Singaporeans: 1995–1999*. The data were collected as part of a follow-up to the 1995 *National Survey of Senior Citizens*. In 1995, a representative sample of 4,750 individuals aged fifty-five and above were interviewed about a variety of issues including demographic characteristics, work and retirement, living arrangements and intergenerational support, income and assets, health status and behaviours, and involvement in voluntary activities and organizations. The 1995 survey over-sampled individuals aged seventy-seven and above, and weights were constructed to account for this.

In 1999, National University of Singapore (NUS) researchers attempted to re-interview as many of the original respondents as possible. This NUS-funded research project was a collaborative effort between researchers at the NUS, the Ministry of Community Development and Sports (Singapore), and the Population Studies Center of the University of Michigan (USA). Although the initial 1995 survey was not designed as a longitudinal study, taking into account the mortality rate for this age group (4 per cent per year) and other losses to follow up (including moves and severe health impairments impeding interview), we re-contacted 42 per cent of the original respondents. This resulted in a sample size of 1,977 older Singaporeans (fifty-nine and above). In order to adjust for panel attrition we calculated a weight adjustment based on the estimated probability of non-response derived from a multivariate regression model including age, sex, ethnicity and marital status as the predictors. The original (baseline) sample weight was multiplied by this weight adjustment in order to obtain the panel sample weight. The adjusted weight was then normalized to ensure that the weighted sample size was equal to the actual sample size, by dividing by its mean. (Full details of the weighting procedure and model specification may be found in Ofstedal *et al.* forthcoming.) We omitted sixteen cases due to missing data and the resulting sample size was 1,961.

Dependent variable

The dependent variable in this study is self-reported functional disability. Functional disability is characterized in terms of a mixture of Nagi indicators, activities of daily living (ADLs) and instrumental activities of daily living (IADLs). Essentially we wanted to construct an indicator showing whether one has difficulty performing any of these actions:

- Preparing own meals
- Bathing, feeding, toileting (ADLs)
- Shopping for groceries/personal needs

- Managing own money
- Doing light housework (e.g. cleaning dishes, straightening up, light cleaning)
- Using transport to get to places that are beyond walking distance
- Crouching or squatting
- Lifting or carrying something as heavy as a 5 kg bag of rice
- Walking 200–300 metres
- Going up and down the stairs (about one or two flights)
- Using fingers to grasp a handle.

Respondents who reported one or more of these limitations were coded as having a functional disability (coded: yes = 1; no = 0).

Independent variables

The main independent variable for this study is nativity defined as native-born and those born in China or Malaysia or elsewhere (India, Indonesia, and others). Consistent with past research that has examined disability (Lawrence and Jette 1996; Kington and Smith 1997; Peek *et al.* 1997, 2003; Rudkin *et al.* 1997; Jette *et al.* 1998; Ostir *et al.* 1998; Zsembik *et al.* 2000; Femia *et al.* 2001; Reynolds and Silverstein 2003), we controlled for several risk factors that are hypothesized to be associated with disability or the disablement process (through their influence on pathology, impairments and functional limitations). Our demographic risk factors included age (fifty-five to sixty-nine, seventy to seventy-nine and eighty and above), sex (male, female), ethnicity (Chinese, Indian, Malay) and marital status (currently married, widowed/divorced/separated, never married). Consistent with previous research, we expected that being older, female, and widowed, divorced or separated, and never married, and belonging to a minority ethnic group would be associated with a higher probability of reporting a functional disability (Angel *et al.* 1996; Hazuda and Espino 1997; Jette *et al.* 1998).

Our socio-economic variables were level of education (none, primary, secondary and above), average monthly income of the respondent and the spouse (less than S\$1,000, S\$1,000+), and adequacy of income (yes, no). Because individuals with higher socio-economic status (SES) have the knowledge, resources and social connections to avoid risk or to minimize the effects of disease, impairment and disability (Link and Phelan 2000), we hypothesized an inverse relationship between SES and functional disability.

Our health risk behaviour variables included smoking (never smoked, quit smoking, currently smoking) and drinking alcohol (non-drinker, former drinker, current drinker). There is strong evidence that current or former smoking is a risk factor for functional status decline (Branch 1985; House *et al.* 1994; Liu *et al.* 1995; Stuck *et al.* 1999). However, the health implications of alcohol consumption are complex because regular low/moderate alcohol consumption has been associated with a decreased risk of heart disease (Stuck *et al.* 1999), while excessive drinking has serious adverse effects (WHO 2001). Moreover, due to the low prevalence of heavy drinking among women, this association was only significant among men (LaCroix *et al.* 1993). We hypothesized that individuals who smoked and drank alcohol would be more likely to report a functional ability.

We also included an indicator of perceived health status which is dichotomous; poor self-assessed health (1) and good self-assessed health (0).[1] Subjective health has been found to be independently associated with functional disability, with greater risk of disability among those with fair or poor self-rating, compared to those with good self-rating (Goldman *et al.* 1995; Idler and Kasl 1995; Femia *et al.* 2001). We hypothesized that individuals who rated their health as poor would be more likely to report a functional ability. However, we did not rule out the possibility of a reverse association: that having a functional disability could result in one's poor self-assessment of health.

Our social networking variables included availability of emotional support (very little, some, quite good), organizational membership (yes, no), living arrangements (alone versus with spouse/children/in-laws/others). Consistent with previous research, we hypothesized that perceived emotional support would be associated with lower disability (see e.g. Thoits (1995) for a review of the effects of various dimensions of social support on health). In addition, we hypothesized that individuals with organizational membership and living alone would be associated with a lower risk of reporting a functional disability.

Statistical analysis

Given the dichotomous nature of the dependent variable (0,1), binary-logistic regression was used to estimate univariate as well as multivariate models. Coefficients were estimated using the maximum likelihood method (MLM) of estimation. The independent variables were recoded into categorical indicator variables. One value of each variable was chosen as the reference category. For ease of interpretation, the results are discussed in terms of the odds ratios. The odds ratio is a measure that approximates how much more likely (or unlikely) it is for the outcome, in this case being functionally disabled, to be present among those with a given attribute, relative to the reference category. Both univariate and multivariate models were fitted. The results in the univariate model describe the gross effect; while the results in the multivariate model describe the net effect, that is, the effect of nativity after controlling for the effects of other variables in the model. A variable was considered significantly associated with the probability of the presence of a functional disability when its *p* value was below 0.05.

We started with a model including only nativity as an independent variable to examine the gross effect of nativity on functional disability. Second, a multivariate model was fitted to see the net effect of nativity on functional disability after simultaneously controlling for demographic, socio-economic, health behaviour, current self-assessed health status and social networking factors.

Results

The characteristics of the sample are shown separately for the native-born and foreign-born in China, Malaysia and elsewhere in Table 5.1. These data were weighted as described above. Sixty per cent of the sample consisted of native-born. The native-born and foreign-born from Malaysia have, by and large, similar characteristics. For example, 27 per cent and 43 per cent of the native-born and 29 per cent and 46 per cent of the foreign-born in Malaysia reported a disability and poor self-assessed health, respectively. The respective proportions are 46 per cent and 50 per cent for the China-born, and 40 per cent and 60 per cent for those born elsewhere.

The average age of the foreign-born from China and those born elsewhere was older; 77 per cent of the foreign-born in China and 46 per cent of the foreign-born elsewhere were in the seventy years and above age group, compared to 31 per cent of native-born and only 24 per cent of those born in Malaysia. This accounts for higher functional disability among foreign-born in China and those born elsewhere. The three main ethnic groups were represented in the sample. Chinese formed the majority ethnic group (87 per cent of older native-born and 79 per cent of older foreign-born), while Indians and Malays formed the minority groups. While the native-born, Malaysia-born and China-born groups were dominated by females, the reverse was true for those born elsewhere. While more than 50 per cent of the native-born, Malaysia-born and those born elsewhere were currently married, only 43 per cent of those born in China were currently married.

Native-born and foreign-born older adults differed substantially in their socio-economic characteristics. Individuals born elsewhere were more likely to have completed some form of education compared to the native-born; 45 per cent of older foreign-born elsewhere were uneducated compared to 58 per cent of older native-born adults, 57 per cent of those born in Malaysia and 73 per cent of those born in China. The older native-born, and those born in Malaysia, were more likely to have monthly incomes of above S$1,000 per month, compared to older foreign-born from China or elsewhere; 81 per cent of older foreign-born from China and 71 per cent of foreign-born elsewhere have incomes of below S$1,000 per month, compared to 67 per cent of native-born and 63 per cent of foreign-born in Malaysia.[2] However, the foreign-born were more likely to report their income as adequate, compared to the native-born; 85 per cent, 82 per cent and 91 per cent of older foreign-born from China, Malaysia and elsewhere, reported respectively that they had an adequate income, compared to 81 per cent of the native-born. There were no significant differences by nativity in terms of smoking behaviour, but there were some differences in terms of drinking behaviour, with a higher proportion of never-drinkers among those born in Malaysia, compared to any other group.

Older foreign-born were more likely to participate in organizational membership compared to older native-born. Foreign-born from Malaysia or elsewhere reported quite good emotional support from family and friends compared to older native-born or those foreign-born in China. A large majority of older adults, both native-born and foreign-born in Singapore, lived with a spouse and/or a child, and this reflects traditional attitudes towards living arrangement patterns and high housing costs (Mehta *et al.* 1992; Chan 1997).

A univariate logistic regression model is presented in Table 5.2. We found that there is a significant difference between the native-born and foreign-born in functional disability. The odds ratios indicate that, compared to native-born older adults, those born in China or elsewhere were more than twice as likely to report being functionally disabled. However, there was no difference in reporting a functional disability between the native-born and the foreign-born from Malaysia. To assess the relative effect of nativity and selected characteristics on self-reported functional disability among older adults in Singapore, we controlled for all the risk factors simultaneously. In this multivariate model (Table 5.3), there was no difference in functional disability between the native- and foreign-born. In fact, the nativity differences in health disappear once demographic factors are controlled for (see Table 5.4). The addition of socio-economic, health risk factors, current self-assessed health and social networking variables to the model already containing demographic factors does not change the results.

Since the native-born and foreign-born difference in functional disability disappears when demographic factors are introduced into the model, we conclude that the nativity disparity in functional disability is due primarily to demographic factors. That is, all of the observed excess in functional

Table 5.2 Summary results from logistic regression model (gross model) for the effect of nativity on functional disability among older adults, Singapore, 1999

Variable	Regression coefficient (β)	Exp (β)	SE	95% confidence interval
Nativity				
Native-born	1.000	1.000	1.000	1.000
China-born	0.758	2.134 (0.000)	0.104	1.741–2.616
Malaysia-born	0.036	1.037 (0.811)	0.151	0.772–1.392
Others-born	0.769	2.158 (0.001)	0.230	1.374–3.390

Source: *Transitions in Health, Wealth, and Welfare of Elderly Singaporeans: 1995–1999* survey.

Note
Figure in parenthesis is actual significance level.

Table 5.3 Odds ratios showing the effects of nativity, demographic, socio-economic, health risks behaviours and social networking variables on functional disability among older Singaporean adults, total sample

Variables	Regression coefficient (β)	Exp (β)	SE	95% confidence interval
Nativity				
Native-born	0.000	1.000		
China-born	0.109	1.115	0.140	0.847–1.468
Malaysia-born	0.020	1.021	0.186	0.709–1.470
Born elsewhere	0.169	1.184	0.313	0.641–2.186
Age				
Adults aged 55–69	0.000	1.000		
Adults aged 70–79	0.636	1.888***	0.144	1.424–2.504
Adults 80 and over	1.527	4.608***	0.165	3.330–6.362
Sex				
Male	0.000	1.000		
Female	0.649	1.913***	0.158	1.404–2.607
Ethnicity				
Chinese	0.000	1.000		
Indian	0.579	1.785	0.327	0.940–3.388
Malay	0.455	1.577**	0.202	1.061–2.342
Marital status				
Currently married	0.000	1.000		
Widowed/divorced/separated	0.229	1.257	0.141	0.954–1.657
Never married	0.031	1.032	0.361	0.508–2.093
Highest education level achieved				
Completed secondary and above	0.000	1.000		
Completed primary	0.157	1.170	0.230	0.746–1.835
No education	−0.099	0.906	0.221	0.588–1.396
Average monthly income of respondent and spouse				
More than S$1,000	0.000	1.000		
Less than S$1,000 (including none)	0.289	1.335	0.154	0.988–1.805
Adequacy of income				
Yes	0.000	1.000		
No	0.238	1.269	0.146	0.954–1.688
Smoking				
Never smoked	0.000	1.000		
Quit smoking	−0.087	0.917	0.159	0.671–1.253
Current smoker	−0.405	0.667**	0.185	0.464–0.957

Table 5.3 continued

Variables	Regression coefficient (β)	Exp (β)	SE	95% confidence interval
Alcohol consumption				
Abstain from drinking/never consumed alcohol	0.000	1.000		
Stopped drinking	0.156	1.168	0.201	0.787–1.734
Currently consuming alcohol	−0.159	0.853	0.205	0.570–1.276
Emotional support				
Quite good	0.000	1.000		
Some	−0.613	0.542***	0.165	0.392–0.748
Very little/not at all	−0.318	0.727**	0.129	0.565–0.936
Organisation membership				
Yes	0.000	1.000		
No	−0.183	0.833	0.182	0.583–1.190
Living arrangements				
With spouse/children/in-laws/ other persons	0.000	1.000		
Alone	−0.546	0.579**	0.244	0.359–0.935
Poor self-assessed health				
No	0.000	1.000		
Yes	1.721	5.592***	0.115	4.460–7.010

Source: *Transitions in Health, Wealth, and Welfare of Elderly Singaporeans: 1995–1999* survey.

Note
*** $p < 0.001$;
** $p < 0.05$.

disability among the foreign-born from China or those born elsewhere as demonstrated in the gross model (see Table 5.2) is accounted for by the demographic factors included in the final multivariate model (Table 5.3). The results demonstrate that the less favourable distribution of age composition of foreign-born from China and those born elsewhere does account for their higher functional disability *vis-à-vis* native-born or foreign-born from Malaysia. Among our demographic risk factors, older adults aged seventy or older, and females, were more likely to report at least one functional disability relative to older adults aged fifty-nine to sixty-nine and males. Our results are similar to Peek *et al.* (2003), who found that being older and female was associated with increased lower body functional limitations.

None of the socio-economic or health risk factors have a significant effect on the probability of reporting a functional disability. Rather, social networking and self-assessed health status significantly affect the self-reporting of functional disability. Older adults who have either some emotional support or no emotional support were less likely to report functional

Table 5.4 Odds ratios showing the effects of nativity and demographic variables on functional disability among older Singaporean adults, total sample

Variables	Regression coefficient (β)	Exp (β)	SE	95% confidence interval
Nativity				
Native-born	0.000	1.000		
China-born	0.135	1.144	0.128	0.890–1.470
Malaysia-born	0.001	1.001	0.170	0.718–1.396
Born elsewhere	0.222	1.248	0.280	0.721–2.162
Age				
Adults aged 55–69	0.000	1.000		
Adults aged 70–79	0.705	2.023***	0.130	1.567–2.612
Adults 80 and over	1.695	5.447***	0.147	4.086–7.262
Sex				
Male	0.000	1.000		
Female	0.836	2.307***	0.122	1.816–2.930
Ethnicity				
Chinese	0.000	1.000		
Indian	1.152	3.163***	0.294	1.778–5.628
Malay	0.860	2.363***	0.174	1.680–3.323
Marital status				
Currently married	0.000	1.000		
Widowed/divorced/separated	0.073	1.075	0.124	0.843–1.371
Never married	−0.294	0.746	0.325	0.395–1.409

Source: *Transitions in Health, Wealth, and Welfare of Elderly Singaporeans: 1995–1999* survey.

Note
*** $p < 0.001$.

disability, compared to those who had quite good emotional support. It may be that individuals who have a functional disability are more in need of emotional support and, therefore, receive it and/or live with somebody. Our results are inconsistent with Peek *et al.* (2003), who found a higher level of emotional support associated with a decrease in lower body functional limitation, and they reasoned that the support network 'rallies' during times of need. Thus an increase in need could be associated with an increase in support (Peek *et al.* 2003).

We found that older adults who live alone were less likely to report a functional disability, but this may be a result of reverse causality whereby older adults who live alone do so because they are healthier. There is a positive and significant correlation between self-assessed health and the likelihood of reporting a functional disability. Respondents who reported their self-assessed health as poor were almost six times more likely to report a functional disability.

Discussion

The aim of this study was to analyse the influence of country of birth (Singapore versus those born in China, Malaysia or elsewhere) on the probability of reporting a functional disability among older Singaporeans. We wanted to explore whether nativity is an independent risk factor for reporting a functional disability, or whether nativity differences in self-reported functional disability merely reflect differences in demographic and socio-economic characteristics. The results indicate that nativity is not a significant risk factor for reporting functional disability once demographic factors are controlled for (Table 5.4). Thus the difference in functional disability between native- and foreign-born older adults is largely explained by the demographic variables in our model. Socio-economic, health behaviour and social networking factors did not affect the nativity differences in functional disability to any significant degree.

How do we explain the absence of nativity differences in functional health among older adults in Singapore, given the significant amount of literature, particularly in the USA and Europe, that demonstrates nativity differences in health, despite controlling for all the confounders? Our interpretation lies in the decline of the healthy migration effect with ageing. We make the assumption that foreign-born older adults were very healthy people at the time of arrival in Singapore. We make this assumption because the strong and healthy are capable of migrating (Sundquist *et al.* 2003), healthy people can be deliberately selected in the immigration process, and/or the healthy workers are recruited by the companies hiring them. However, the healthy migrant effect's influence on health decreases with time (Sundquist and Johansson 1997a). One possible reason for the decline in the healthy migrant effect may be that psycho-social and economic conditions in the new country negatively affect health (Sundquist *et al.* 2003). In another Swedish national study that focused on Iranians, Chileans, Poles, Turks and Kurds, for example, economic difficulties in their new country and poor acculturation (men only) were stronger risk factors for psychological distress than exposure to violence before migration (Sundquist *et al.* 2000).

Indeed, a number of studies have shown that although immigrants are generally healthy at the time of arrival, their health status (measured through self-assessed health or other measures including mental health) tends to converge (downward) towards the host population over time (Chen *et al.* 1996; Dunn and Dyck 2000; Frisbie *et al.* 2001; Muenning and Fahs 2002). Perez (2002), for example, noted that the likelihood of reporting any chronic condition among immigrants increased with time spent in Canada, despite initially superior health relative to the Canadian-born. In another study, Newbold and Danforth (2003) found a near continuous decline in health status among immigrants with increasing duration of residence, and they attributed this difference to the ageing of the immigrant population: 'increasing duration of residency within Canada is directly linked with age'

(Newbold and Danforth 2003: 1985). Thus the absence of differences in functional disability among foreign-born and native-born may be influenced by the ageing of the foreign-born and the decline of the 'healthy migrant effect' with time (Williams 1993; Johansson *et al.* 1997). Since we found no significant differences in the functional disability between native-born and foreign-born, it is reasonable to assume that general efforts to lower functional disability will be beneficial for all groups.

Limitations

There are several limitations to this study. First, it is based on a cross-sectional survey design that does not allow one to draw inferences about causal pathways. Second, the definition of functional disability was limited to reporting any functional disability versus no disability. This is a crude definition, since those with a functional disability are not homogeneous in terms of severity of disability. Third, the use of self-reported data may introduce some reporting bias whereby some groups (for example, the less educated and older people) may be less likely to accurately recall health conditions. Vargas *et al.* (1997) found that men (particularly non-whites) and people who had not received medical care in the previous year showed the least correspondence between self-reported measures and hypertension. Other researchers have expressed the opinion that self-rated data are too subjective; that is, that they are not as objective a measure of health status as are standardized mortality rates or diagnoses obtained from health examinations (Sundquist *et al.* 2003). Nevertheless, studies in the USA and Europe have shown that self-reported health status is, in fact, a strong predictor of mortality (Sundquist and Johansson 1997b; McGee *et al.* 1999). Finally, the results from this study cannot be generalized to other contexts. However, in the context of Singapore, the findings in this study are important to policymakers and planners in terms of providing for elderly care: government policies to promote health care for all should apply to all citizens, including members of the foreign-born group.

Acknowledgements

Funding for this research was provided by a United States National Institute of Aging grant, *Comparative Study of Health Transitions among Older Adults in Asia* and by a National University of Singapore sponsored grant, R-111–000–039–593, *Transitions in Health, Wealth, and Welfare among Singaporean Elderly: 1995–1999*. We have benefited from comments on an earlier draft by Professor Paul Boyle and would like to thank him.

Notes

1 A number of studies have also examined the association between chronic disease and functional limitation (see e.g. Boult *et al.* (1994) for a review of findings demonstrating a relationship among various chronic health conditions and increases in functional limitations), and suggested that chronic conditions are predictive of functional limitations (Markides *et al.* 1996; Perkowski *et al.* 1997; Ma *et al.* 1998). In this study we did not include this variable for two main reasons. First, everyone in the sample who reported functional disability recorded themselves as having a chronic condition. Second, inclusion or exclusion of this variable did not change the effect of nativity on functional disability.
2 At the time of writing (July 2004), US$1 was equivalent to S$1.77.

References

Alston, L.T. and Aguirre, B. (1987) 'Elderly Mexican Americans: nativity and health access', *International Migration Review,* 21, 3: 626–42.

Angel, J.L., Angel, R.J., McClellan, J.L. and Markides, K.S. (1996) 'Nativity, declining health, and preferences in living arrangements among elderly Mexican Americans: implications for long-term care', *The Gerontologist,* 36, 464–73.

Applegate, W.B., Blass, J.P. and Williams, T.F. (1990) 'Instruments for the functional assessment of older patients', *New England Journal of Medicine,* 322, 1207–14.

Axen, E. and Lindstrom, M. (2002) 'Ethnic differences in self-reported lack of access to a regular doctor: a population-based study', *Ethnicity and Health,* 73, 3: 195–207.

Bayard, M. (1978). 'Ethnic identity and stress: the significance of sociocultural context', in J.M. Casas and S.E. Keefe (eds) *Family and Medical Health in the Mexican American Community,* Monograph No. 7, Spanish Speaking Mental Health Research Centre, UCLA, Los Angeles, pp. 109–23.

Bayard-Burfield, L., Sundquist, J. and Johansson, S.E. (1998) 'Self-reported long-standing psychiatric illness as a predictor of premature all-cause mortality and violent death: a 14-year follow-up study of native Swedes and foreign-born immigrants', *Social Psychiatry and Psychiatric Epidemiology,* 33: 491–6.

Bayard-Burfield, L., Sundquist, J., Johansson, S.E. and Traskman-Bendz, L. (1999) 'Attempted suicide among Swedish-born people and foreign-born immigrants', *Archives of Suicide Research,* 5: 43–55.

Boult, C., Kane, R.L., Louis, T.A., Bouldt, L. and McCaffrey, D. (1994) 'Chronic conditions that lead to functional limitation in the elderly', *Journal of Gerontology: Medical Sciences,* 49A: M28–36.

Bowling, A., Farquar, M. and Leaver, J. (1992) 'Jewish people and ageing: their emotional well-being, physical health status and use of services', *Nursing Practice,* 5: 5–16.

Branch, L.G. (1985) 'Health practices and incident disability among the elderly', *American Journal of Public Health,* 75: 1436–9.

Chan, A. (1997) 'An overview of the living arrangements and social support exchanges of older Singaporeans', *Asia-Pacific Population Journal,* 12, 4: 35–50.

Chan, C-B. and Tong, C-K. (eds) (2003) *Past Times: A Social History of Singapore,* Singapore: Times Editions.

Chen, J., Wilkens, R. and Ng, E. (1996) 'Life expectancy of Canada's immigrants from 1986 to 1991', *Health Reports,* 8, 3: 29–38.

Chung, R.C.Y. and Kagawa-Singer, M. (1993) 'Predictors of psychological distress among Southeast Asian refugees', *Social Science and Medicine,* 36: 631–9.

Department of Statistics (2004) online, available at: www.singstat.gov.sg (accessed 21 September 2004).

Dunn, J. and Dyck, I. (2000) 'Social determinants of health in Canada's immigrant population: results from the national population health survey', *Social Science and Medicine,* 51: 1573–93.

Erb, M. (2003). 'Moulding a nation: education in early Singapore', in C.K. Tong and K.B. Chan (eds) *Past Times: A Social History of Singapore,* Singapore: Times Editions.

Femia, E.E., Zarit, S.H. and Johansson, B. (2001) 'The disablement process in very late life: a study of the oldest old in Sweden', *Journal of Gerontology: Psychological Sciences,* 56B: P12–23.

Fried, L.P. and Guralnik, J.M. (1997) 'Disability in older adults: evidence regarding significance, etiology and risk', *Journal of the American Geriatrics Society,* 45: 92–100.

Frisbie, W.P., Youngtae, C. and Hummer, R.A. (2001) 'Immigration and the health of Asian and Pacific Islander adults in the United States', *American Journal of Epidemiology,* 153, 4: 372–80.

Goldman, N., Korenman, S. and Weinstein, R. (1995) 'Marital status and health among the elderly', *Social Science and Medicine,* 40: 1717–30.

Hazuda, H.P. and Espino, D. (1997) 'Aging, chronic disease, and physical disability in Hispanic elderly', in K.S. Markides and M. Miranda (eds) *Minorities, Aging, and Health,* Newbury Park, CA: Sage Publications.

House, J.S., Lepkowski, J.M., Kinney, A.M., Mero, R.P., Kessler, R.C. and Herzog, R.A. (1994) 'The social stratification of aging and health', *Journal of Health and Social Behaviour,* 35: 213–34.

Idler, E.L. and Kasl, S.V. (1995) 'Self-ratings of health: do they also predict change in functional ability?', *Journal of Gerontology: Social Sciences,* 50: S344–53.

Jette, A.M., Assmann, S.F., Rooks, D., Harris, B.A. and Crawford, S. (1998) 'Inter-relationships among disablement concepts', *Journal of Gerontology: Medical Sciences,* 53A, M395–404.

Johansson, L.M., Sundquist, J., Johansson, S.E. and Bergman, B. (1997) 'The influence of ethnicity and social and demographic factors on Swedish suicide rates: a four year follow-up study', *Social Psychiatry and Psychiatric Epidemiology,* 32: 165–70.

Kaplan, M. and Marks, G. (1990) 'Adverse effects of acculturation: psychological distress among Mexican-American young adults', *Social Science and Medicine,* 31: 1313–19.

Keefe, S.E. (1982) 'Help-seeking behaviour among foreign-born and native-born Mexican Americans', *Social Science and Medicine,* 16: 1467–74.

Keefe, S.E., Pidilla, A.M. and Carlos, M.L. (1979) 'The Mexican American extended family as an emotional support system', *Human Organisation,* 38 (spring): 144–52.

Kindlund, H. (1995) 'Early retirement pension and sick leave among immigrants and Swedes in 1990', *Socialstyrelsen rapport,* 5: 135–51.

Kington, R.S. and Smith, J.P. (1997) 'Socioeconomic status and racial and ethnic differences in functional status associated with chronic diseases', *American Journal of Public Health,* 87, 5: 805–10.

Kuo, W.H. and Tsai, Y. (1986) 'Social networking, hardiness and immigrant's mental health', *Journal of Health and Social Behaviour,* 27: 133–49.

LaCroix, A.Z., Guralnik, J.M., Berkman, L.F., Wallace, R.B. and Satterfield, S. (1993) 'Maintaining mobility in late life: II Smoking, alcohol consumption, physical activity and body mass index', *American Journal of Epidemiology,* 137: 858–69.

Lamb, V.L. (1997) 'Gender differences in correlates of disablement among the elderly in Egypt', *Social Science and Medicine,* 45, 1: 127–36.

Lawrence, R.H. and Jette, A.M. (1996) 'Disentangling the disablement process', *Journal of Gerontology: Social Sciences,* 51B: S173–82.

Link, B.G. and Phelan, J.C. (2000) 'Evaluating the fundamental cause explanation for social disparities in health', in C.E. Bird, P. Conrad and A.M. Fremont (eds) *Handbook of Medical Sociology,* Upper Saddle River, NJ: Prentice Hall, pp. 33–46.

Liu, X., Liang, J., Muramatsu, N. and Sugisawa, H. (1995) 'Transitions in functional status and active life expectancy among older people in Japan', *Journal of Gerontology: Social Sciences,* 50, S383–94.

Ma, J., Markides, K.S., Stroup-Benham, C.A., Lichtenstein, M. and Goodwin, J.S. (1998) 'Impact of selected medical conditions on lower-extremity function in Mexican American elderly', *Ethnicity and Disease,* 8: 52–9.

McGee, D.L., Liao, Y., Cao, G. and Cooper, R.S. (1999) 'Self-reported health status and mortality in a multiethnic US cohort', *American Journal of Epidemiology,* 149: 42–6.

Markides, K.S., Stroup-Benham, C.A., Goodwin, J.S., Perkowski, L.C., Lichtenstein, M. and Ray, L.A. (1996) 'The effect of medical conditions on the functional limitations of Mexican-American elderly', *Annals of Epidemiology,* 6: 386.

Marmot, M.G., Adelstein, M.A. and Bulusu, L. (1984) 'Immigrant mortality in England and Wales 1970–78', OPCS Studies of Medical and Population Subjects No. 47, London: HMSO.

Mayhew, L. (1999) *Health and Welfare Services Expenditure in an Aging World,* Interim Report. International Institute for Applied Systems Analysis, Number IR-99–035/September.

Mehta, K., Lee, A.E.Y. and Osman, M. (1992) 'Living arrangements of the elderly in Singapore: cultural norms in transition', *Comparative Study of the Elderly in Asia Research Report 92–22,* Population Studies Center, University of Michigan, Ann Arbor.

Muenning, P. and Fahs, M.C. (2002) 'Health status and hospital utilization of recent immigrants to New York City', *Preventive Medicine,* 35: 225–31.

Newbold, K.B. and Danforth, J. (2003) 'Health status and Canada's immigrant population', *Social Science and Medicine,* 57: 1981–95.

Ofstedal, M.B., Agree, E., Chan, A., Chuang, Y-L., Costenbader, E., Kaneda, T., Natividad, J., Zhe T. and Zimmer, Z. (forthcoming) 'Analysis of sample attrition in the aging and health in Asia surveys', PSC Elderly in Asia Reseach Report, Population Studies Center, University of Michigan, Ann Arbor.

Ostir, G., Markides, K.S., Black, S.A. and Goodwin, J.S. (1998) 'Lower body functioning as a predictor of subsequent disability among older Mexican Americans', *Journal of Gerontology: Medical Sciences,* 54A: M491–5.

Peek, C.W., Coward, R.T., Henretta, J.C., Duncan, R.P. and Dougherty, M.C. (1997) 'Differences by race in the decline of health over time', *Journal of Gerontology: social sciences,* 52B: S336–44.

Peek, C.W., Ottenbacher, K.J., Markides, K.S. and Ostir, G.V. (2003) 'Examining the disablement process among older Mexican American adults', *Social Science and Medicine,* 57: 413–25.

Perez, C.E. (2002) 'Health status and health behaviour among immigrants', *Supplement to Health Reports*, Statistics Canada, Catalogue 82–003–SIE 112.

Perkowski, L.C., Stroup-Benham, C.A., Markides, K.S., Lichtenstein, M.J., Angel, R.J. and Goodwin, J.S. (1997) 'The association of medical problems with performance based measures of lower extremity functioning in older Mexican Americans', *Journal of American Geriatrics Society*, 46: 411–18.

Raftery, J., Jones, D. and Rosato, M. (1990) 'The mortality of first and second generation Irish immigrants in the UK', *Social Science and Medicine*, 31: 577–84.

Reynolds, S.A. and Silverstein, M. (2003) 'Observing the onset of disability in older adults', *Social Science and Medicine*, 57: 1875–89.

Rudkin, L., Markides, K.S. and Espino, D.V. (1997) 'Functional disability in older Mexican Americans', *Topics in Geriatric Rehabilitation*, 12: 38–46.

Shuval, J. (1993) 'Migration and stress', in L. Goldberger and S. Breznitz (eds) *Handbook of Stress: Theoretical and Clinical Aspects* (2nd edn), New York: The Free Press, pp. 677–91.

Stephenson, P.H. (1995) 'Vietnamese refugees in Victoria, BC: an overview of immigrant and refugee health care in a medium-sized urban centre', *Social Science and Medicine*, 40: 1631–42.

Stuck, A.E., Walthert, J.M., Nikolaus, T., Bula, C.J., Hohmann, C. and Beck, J.C. (1999) 'Risk factors for functional status decline in community living elderly people: a systematic literature review', *Social Science and Medicine*, 48: 445–69.

Sundquist, J. (1994) 'Refugees, labour migrants and psychological distress', *Social Psychiatric Epidemiology*, 29: 20–4.

Sundquist, J. (1995) 'Ethnicity, social class and health: a population-based study of social factors influence on self-reported long-term illness, illness incidence, working impairment and disability in 223 Latin American refugees, 333 Finnish and 126 South European labour migrants and 841 sex, age- and education-matched Swedish controls', *Social Science and Medicine*, 40, 6: 777–87.

Sundquist, J. and Johansson, S.E. (1997a) 'Long-term illness among indigenous and foreign-born people in Sweden', *Social Science and Medicine*, 44: 189–98.

Sundquist, J. and Johansson, S.E. (1997b) 'Self reported poor health and low educational level predictors for mortality: a population based follow-up study of 39,156 people in Sweden', *Journal of Epidemiology and Community Health*, 51: 35–40.

Sundquist, J., Bayard-Burfield, L., Johansson, L.M. and Johansson, S.E. (2000) 'Impact of ethnicity, violence and acculturation on displaced migrants: psychological distress and psychosomatic complaints among refugees in Sweden', *Journal of Nervous and Mental Disease*, 188: 357–65.

Sundquist, J., Ostergren, P.O., Sundquist, K. and Johansson, S.E. (2003) 'Psychosocial working conditions and self-reported long-term illness: a population-based study of Swedish-born and foreign-born employed persons', *Ethnicity and Health*, 8, 4: 307–17.

Thoits, P. (1995), 'Stress, coping and social support processes: where are we? What next?', *Journal of Health and Social Behaviour*, Extra Issue: 53–79.

Vargas, C.M., Burt, V.L., Gillum, R.F. and Pamuk, E.R. (1997) 'Validity of self-reported hypertension in the National Health and Nutrition Examination Survey III, 1988–1991', *Preventive Medicine*, 26: 678–85.

Verbrugge, L.M. and Jette, A.M. (1994) 'The disablement process', *Social Science and Medicine*, 38: 1–14.

Williams, R. (1993) 'Health and length of residence among South Asians in Glasgow: a study controlling for age', *Journal of Public Health Medicine,* 15: 52–60.

WHO (2001) *Health for All Database January* 2001, Geneva: WHO.

Zimmer, Z, Martin, L. and Chang, M.C. (2002) 'Changes in functional limitation and survival among older Taiwanese, 1933, 1966 and 1999', *Population Studies,* 3: 265–76.

Zimmer, Z., Liu, X., Hermalin, A.I. and Chuang, Y.L. (1998) 'Educational attainment and transition in functional status among older Taiwanese', *Demography,* 35, 3: 361–75.

Zsembik, B.A., Peek, M.K. and Peek, C.W. (2000) 'Race and ethnic variation in the disablement process', *Journal of Aging and Health,* 12: 229–49.

6 Anaemia among migrant and non-migrant mothers in disadvantaged areas in the Visayas, the Philippines

Alan Banzon Feranil

Introduction

This chapter examines the anaemia status of migrant and non-migrant mothers in disadvantaged areas of the Visayas, one of the three major island groups in the Philippines. The factors associated with maternal anaemia among mothers are also explored in this chapter, taking into consideration individual, household and community characteristics.

Migration is emerging as a major health concern for several reasons. There is a growing concern about the health of migrants because of the impact of migration on the spread of diseases such as acute immunodeficiency syndrome, sexually transmitted diseases (Duckett 2000; Gupta and Mitra 1999), child survival (Sastry *et al.* 1993) and the availability of health services for migrants (Chantavanich and Paul 1999; Littlefield and Stout 1987; Teller 1973). In some situations, the places of destination do not have the capacity to absorb the needs of migrants. However, migration in some cases may not result in an unmet health need since migration may be a selective process, preferring healthier and more affluent individuals who have the resources to move to a better environment.

Studies on health and migration have shown migrant status to be an important attribute that may be associated with health status in populations. While some studies have demonstrated that migrants are disadvantaged in their places of destination with respect to their health status (e.g. Bollini and Siem 1995: 823), other studies have shown that migrant populations tend to be healthier, with lower mortality rates and higher life expectancies than the native born populations (e.g. Muennig and Fahs 2002: 229). Although some studies have indicated that migrants who have been socialized in a more traditional social medical environment and assimilated into an urban environment tend to have similar health behaviour to urban populations (e.g. Teller 1973: 222), other studies have implied otherwise. A study of Turkish residents in Germany, for example, revealed that second-generation migrant settlers still have better health status than the native-born populations (e.g. Razum *et al.* 1998: 299). These studies indicate that migration is an important characteristic of health status.

The health status of migrants is a subject that has not been well explored in the Philippines setting. Most Philippine studies of migration have focused on the demographic pursuits of identifying migration streams and motivating factors (Abejo 1985; Amacher *et al.* 1998; Flieger *et al.* 1976; Gonzales and Pernia 1983; Perez 1978, 1985) and policies (Abella 1989). Other Philippine studies have focused on the overseas migration of Filipino workers (Battistella 1995) or on the plight of female migrants (Asis 1990; Go 1995). Some studies have focused on the effect of migration on the environment (Eder 1990), and social costs and reintegration of female migrants (Dizon-Añonuevo and Añonuevo 2003), rather than on migrant health. Philippine studies that have specifically focused on, or included, migrant health as an issue of concern are limited. The studies have focused primarily on migrant fertility behaviour (Sembrano 1980) or on particular health problems such as cardiovascular disorders (Cabral *et al.* 1983). This chapter examines another health concern – maternal anaemia.

Anaemia is one of the most common health problems in developing countries, not only among young infants and children, but also among women (WHO 2002: 54). Women, particularly those living in rural and tropical areas, not only have to contend with demanding domestic responsibilities including taking care of the household and earning a living, but are also afflicted with iron deficiency anaemia. These factors contribute to their health situation, and may result in significant losses in productivity and affect family welfare (Sims 1994: 36). As in other developing countries, anaemia persists as one of the major nutritional disorders in the Philippines. Statistics from the National Nutrition Surveys in 1993 and 1998 attest to a persistent anaemia problem, not only among children but also among mothers (Food and Nutrition Research Institute 2001: 72).

Anaemia among Filipino mothers is on the rise, and among pregnant mothers is highly prevalent in twelve of the seventeen regions of the Philippines. In fact, the prevalence of maternal anaemia has risen significantly from 43.6 per cent in 1993 to 50.7 per cent in 1998 among pregnant mothers, and from 43.0 per cent to 45.7 per cent among lactating mothers (Cheong *et al.* 2001: 53). The regions focused on in this chapter have been shown to have high incidences of maternal anaemia compared to the national estimates (Cheong *et al.* 2001: 50).

By focusing on a major nutritional disorder such as anaemia, and on the migrant status of mothers, I seek to shed light on the relationship between maternal anaemia and migration and, by doing so, contribute to the existing fund of knowledge and provide some recommendations for policy implementation. I focus on internal migration, an area which has been overshadowed by the current focus on international and overseas labour force migration.

I seek to address the question of whether being a migrant has a bearing on maternal health, or whether being a migrant means being in a state of worse health than the native-born population, which is supposedly more aware of services that are available in the community and has supposedly adapted to the local setting. I examine this perspective in disadvantaged areas identified to be in need and 'at risk' by the government, and consider whether, in such conditions, the argument that the native-born population is healthier than the migrants applies. I provide a different perspective of migrant health by using a biochemical index such as anaemia which is based on actual haemoglobin-level readings. In addition, I provide insights into internal migration by examining the health status of both the native-born and migrant populations.

Methods

This study uses data collected from the 2001 Early Childhood Development (ECD) Program's Baseline Indicators Survey. The ECD Survey included *barangays* (villages) classified to be 'at risk' and in need by the Philippine Government's Department of Social Welfare and Development (Council for the Welfare of Children 1999).[1] The sample mothers in this study refer to those included in the ECD Baseline Indicators Survey for the Early Childhood Development Program, an initiative of the Philippine government in partnership with different line agencies and participating local government units. The Program seeks to address the nutritional, health and psychosocial development of preschool children, and collaborates with local government in the provision of early childhood development packages or intervention programmes supervised by local line agencies of the government, and supported by local, national and international sources. Thus the sample mothers from these *barangays* provide a different insight into the status of migrant and non-migrant mothers living in a disadvantaged setting.

I focus on 4,942 mothers with 0–6-year-old children with complete information on their migration status. Mothers are classified as either native-born, those residing in their place of birth, or movers, those residing in places other than their place of birth. Other migration characteristics, such as length of residence and place of origin, are also explored in relation to their anaemia status. The mothers were drawn from the *barangays* in the Program regions – Western and Central Visayas regions (Regions 6 and 7) – and *barangays* from the control region – Eastern Visayas (Region 8) – included in the ECD Baseline Indicators Survey (see Figure 6.1).[2]

Individual, household and community-level questionnaires were administered to gather the baseline information. Moreover, haemoglobin-level readings from the blood samples collected from mothers by licensed medical technologists were used to ascertain the state of anaemia of mothers included in the study. The classification of mothers as to whether or not they were anaemic was based on the haemoglobin levels set by the World Health Organization.[3]

A key limitation of this research is that the findings are not generalizable to all mothers of preschool children, nor to the municipalities, provinces or

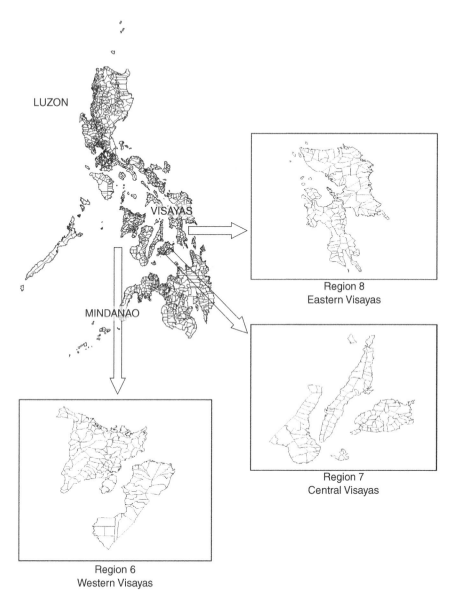

Figure 6.1 Map of the Philippines (source: Office of Population Studies' Geographic Information System, University of San Carlos).

regions from which they are drawn. As mentioned above, the sample *barangays* included in the survey were preselected and do not represent the municipality, province or region to which they belong. The *barangays* were those identified by the Department of Social Welfare and Development to be most in need, and whose local governments were willing to invest in the

Program. Thus the findings are confined to the health status of mothers with children aged 0–6 years old in selected disadvantaged *barangays* in the Visayas Island group. Nevertheless, this sample provides some insights into the association between migration and the anaemia status of mothers living in poor villages, which may differ from the majority of migration studies that often focus on migrant health in urban, more developed or popular areas of destination.

Informed consent was solicited and obtained prior to interview and the drawing of blood samples for haemoglobin analysis.[4] An informed consent form was given to each participant to read, concur with and sign prior to blood collection. Mothers who refused or did not agree to sign were excluded from the blood sample analysis and research data.[5] A list of mothers who were found to be anaemic was sent to the rural health unit for follow-up and appropriate intervention.

Several statistical techniques were used in the analysis. Frequencies and cross-tabulations were used to explore the characteristics of the participants, their migration attributes, and the characteristics associated with anemic and non-anaemic mothers. Chi-square values were used to identify which of the migration characteristics and other individual, household and community characteristics were associated with maternal anaemia. Finally, a binary logistic regression model was carried out using the Statistical Package for Social Sciences (SPSS) software to arrive at a parsimonious model which takes into account the individual, household and community characteristics, and best explains maternal anaemia status.[6]

Results

The characteristics of the sample mothers provide insights into their demographic and socioeconomic attributes which may be important factors associated with their health and migration behaviour. The profile of mothers reveals that more sample mothers tend to be in their thirties, with over half having some high school education (see Table 6.1). Almost all the sample mothers were married and working. On average they had four children which is close to the national average. Only a small proportion of the mothers were pregnant. Although the distribution of the mothers varied across the three regions included in the ECD Baseline Indicators Survey, almost all were from rural areas.

The sample mothers were categorized with respect to four migration attributes: place of birth, place of origin, place of destination, and length of residence. As shown in Table 6.2, a moderate proportion of the sample mothers had moved or were currently residing in an area other than their place of birth. Approximately one-third of the mothers had migrated from rural areas, and more than a quarter came from an urban area (city or *poblacion*). With respect to their length of residence, all the sample mothers had, on average, resided in their current residence for more than seventeen years.

Table 6.1 Characteristics of sample mothers from the Visayas

Characteristics	Percentage	N
Age		
Less than thirty years old	38.6	1,908
At least thirty years old	61.4	3,034
Total	100.0	4,942
Mean age	32.25	95% CI (32.04–32.45)
Education		
Less than high school education	45.6	2,254
At least some high school education	54.4	2,688
Total	100.0	4,942
Marital status		
Currently married	95.5	4,721
Not currently married	4.5	221
Total	100.0	4,942
Work status		
Not working	1.5	68
Working	98.5	4,454
Total	100.0	4,522
Children ever born		
Four children or less	64.3	3,178
More than four children	35.7	1,764
Total	100.0	4,942
Mean children ever born	4.03	95% CI (3.96–4.10)
Pregnancy status		
Currently pregnant	9.9	487
Not currently pregnant	90.1	4,455
Total	100.0	4,942
Residence		
Urban	9.0	447
Rural	91.0	4,495
Total	100.0	4,942
Region		
Region 6	26.9	1,328
Region 7	37.0	1,830
Region 8	36.1	1,784
Total	100.0	4,942

Source: Early Childhood Development Project Baseline Indicators Survey 2001 dataset.

Table 6.2 Migration characteristics of sample mothers from the Visayas

Migration characteristics	Percentage	N
Type of migrant		
Native (non movers)	35.9	1,776
Movers	64.1	3,166
Total	100.0	4,942
Place of origin		
Same *barangay* as birthplace (non-movers)	36.0	1,776
From an urban area (city/*poblacion*)	26.6	1,314
From another rural area	37.4	1,845
Total	100.0	4,935
Mean length of residence in area	17.42	4,942
95% CI	(17.04–17.87)	
Length of residence in area		
Five years or less	27.4	1,354
More than five years	72.6	3,588
Total	100.0	4,942
Length of residence in area		
Ten years or less	43.1	2,132
More than ten years	56.9	2,810
Total	100.0	4,942

Source: Early Childhood Development Project Baseline Indicators Survey 2001 dataset.

Approximately two out of every five mothers had spent ten years or less in their current residence. Hence, more than half of the mothers had spent a considerable amount of time in their place of residence, and had most likely adjusted to the life-style in the area.

A closer analysis of these four migration categories reveals that only the migrant status of being a 'mover' or a native-born mother was statistically significant. As shown in Table 6.3, the association of the mother's migration status and maternal anaemia was statistically significant, with more native-born mothers being anaemic than mover mothers.

With migration status as a characteristic of interest, a description of the characteristics of the native-born and mover mothers reveals that there are more mover mothers who are younger – under thirty years old – and married (see Table 6.4). Although almost all the mothers were living in rural areas, there were slightly more mover mothers in urban areas, and from Regions 7 and 8. With respect to other attributes such as education, current work, pregnancy status and number of children, the chi-square statistics imply no difference between these two groups of mothers.

Maternal anaemia is characterized by the presence of low haemoglobin

Table 6.3 Anaemic status of sample mothers from the Visayas and selected migration characteristics

Characteristics	Percentage			N
	Anaemic	Normal	Total	
Migration status				
Native (non-movers)	36.9	63.1	100.0	1,776
Movers	34.1	65.9	100.0	3,166
Total	–	–	–	4,942
$X^2 = 3.89$, p = 0.048				
Place of origin				
Same *barangay* as birthplace (non-movers)	36.9	63.1	100.0	1,776
From an urban area (city/*poblacion*)	33.3	66.7	100.0	1,314
From another rural area	34.9	65.1	100.0	1,845
Total	–	–	–	4,935
$X^2 = 4.49$, p = 0.106				
Length of residence in area				
Five years or less	34.9	65.1	100.0	1,354
More than five years	35.2	64.8	100.0	3,588
Total	–	–	–	4,942
$X^2 = 0.04$, p = 0.846				
Length of residence in area				
Ten years or less	34.1	65.9	100.0	2,132
More than ten years	35.9	64.1	100.0	2,810
Total	–	–	–	4,942
$X^2 = 1.65$, p = 0.199				

Source: Early Childhood Development Project Baseline Indicators Survey 2001 dataset.

levels and is important, since it increases the risks of maternal mortality in severe cases (Rush 2000: 217S) and may account for significant losses in productivity, especially in developing countries (Sims 1994: 36). Maternal anaemia is also important due to its association with complications such as risks of pre-term delivery (Scholl and Reilly 2000: 444S; Steer 2000: 1286S), low birth weight (Steer 2000: 1286S) and poor child development (Stoltzfus 2001: 698S).

The anaemia status of the sample mothers was explored in relation to particular individual, household and community characteristics. The individual characteristics included age, education, parity, and the taking of iron supplements in the previous past pregnancy.[7] Household characteristics included proxy variables for the household's socioeconomic status and living conditions, such as source of lighting, source of drinking water, type of household toilet, number of individuals in the household, type of household

Table 6.4 Characteristics of sample migrant and native mothers from the Visayas

Characteristics	Percentage			N
	Native	*Movers*	*Total*	
Age				
Less than thirty years old	32.6	67.4	100.0	1,908
At least thirty years old	38.0	62.0	100.0	3,034
Total				4,942
$X^2 = 14.57$, p $= 0.000$				
Mean age	32.72	31.98	32.25	4,942
95% CI	32.38–33.07	31.72–32.23	32.04–32.45	
Education				
Less than high school education	36.4	63.6	100.0	2,254
At least high school education	35.6	64.4	100.0	2,688
Total	–	–	–	4,942
$X^2 = 0.35$, p $= 0.552$				
Marital status				
Currently married	35.4	64.6	100.0	4,721
Not currently married	46.6	53.4	100.0	221
Total	–	–	–	4,942
$X^2 = 11.44$, p $= 0.001$				
Work status				
Not working	29.4	70.6	100.0	68
Working	36.4	63.6	100.0	4,454
Total	–	–	–	4,522
$X^2 = 1.39$, p $= 0.238$				
Children ever born				
Four children or less	35.1	64.9	100.0	3,178
More than four children	37.5	62.5	100.0	1,764
Total	–	–	–	4,942
$X^2 = 2.81$, p $= 0.094$				
Mean children ever born	4.11	3.99	4.03	
95% CI	3.99–4.23	3.90–4.07	3.96–4.10	
Pregnancy status				
Currently pregnant	33.5	66.5	100.0	487
Not currently pregnant	36.2	68.8	100.0	4,455
Total	–	–	–	4,942
$X^2 = 1.43$, p $= 0.232$				

Table 6.4 continued

Characteristics	Percentage			N
	Native	Movers	Total	
Place of residence				
Urban	29.8	70.2	100.0	447
Rural	36.6	63.4	100.0	4,495
Total	–	–	–	4,942
$X^2 = 8.16$, p $= 0.004$				
Region				
Region 6	42.8	57.2	100.0	1,328
Region 7	35.4	64.6	100.0	1,830
Region 8	31.5	68.6	100.0	1,784
Total	–	–	–	4,942
$X^2 = 42.84$, p $= 0.0000$				

Source: Early Childhood Development Project Baseline Indicators Survey 2001 dataset.

composition, and the household head's education. Since environment also plays an important role in health status, proxy variables for the type of community, such as presence of electricity in the neighbourhood and the place of residence (that is, region), were considered.[8]

A binomial logistic regression model (see Model 1, Table 6.5) shows the gross effect of each of the above-mentioned variables. The results show that a mother who is native-born, less educated and who has more children, and who has not taken iron supplements in her previous pregnancy, is predisposed to anaemia. Moreover, living in a nuclear family type of household, with more than six persons, with a non-flush toilet, no lighting, and where the household head has less than a high school education, is associated with maternal anaemia. Similarly, living in a *barangay* with no electricity is also associated with maternal anaemia.

The results of the parsimonious model (see Model 2, Table 6.5) further reveal that maternal anaemia may be best explained by migration status, education, age, parity, the taking of iron supplements and lack of household electricity when all the covariates are taken into consideration. Native-born mothers are more anaemic than mover mothers. Similarly, less educated and younger mothers tend to be more anaemic. Having more children is more closely associated with maternal anaemia, and less with mothers who take iron supplements for their pregnancies. As a proxy of living conditions, living in a household with no lighting is associated with maternal anaemia. Thus, this study has shown that anaemia is also associated with migration. Being a native-born or a mover mother has been significantly associated with the presence or absence of maternal anaemia even when other attributes

Table 6.5 Odd ratios resulting from the logistic regression analysis indicating the influence of selected characteristics on maternal anaemia of sample mothers from the Visayas

Characteristics	Model 1[a]			Model 2[b]		
	Odds ratio	p value	95% CI	Odds ratio	p value	95% CI
Migration status						
Native (non-movers)	1.0000	–	–	1.0000	–	–
Movers	0.8852	0.0486	0.7842–0.9993	0.8661	0.0197	0.7594–0.9764
Education						
Less than high school education	1.0000	–	–	1.0000	–	–
At least high school education	0.7360	0.0000	0.6545–0.8276	0.8368	0.0080	0.7336–0.9545
Age						
Less than thirty years old	1.0000	–	–	1.0000	–	–
At least thirty years old	1.0099	0.8728	0.8957–1.1386	0.8372	0.0147	0.7258–0.9656
Children ever born						
Four children or less	1.0000	–	–	1.0000	–	–
More than four children	1.3949	0.0000	1.22362–1.5739	1.3906	0.0000	1.2003–1.6110
Previous iron supplementation						
Did not take iron supplements	1.0000	–	–	1.0000	–	–
Took iron supplements	0.7071	0.0000	0.6236–0.8017	0.7446	0.0000	0.6543–0.8473
Number of persons in household						
More than six persons	1.0000	–	–	–	–	–
Six persons or less	0.7671	0.0000	0.6804–0.8649	–	–	–
Type of household toilet						
Other types of toilets	1.0000	–	–	–	–	–
Flush or water sealed	0.7721	0.0000	0.6852–0.8700	–	–	–

	Model 1			Model 2		
	OR	p	95% CI	OR	p	95% CI
Type of household						
Extended type	1.0000					
Nuclear type	1.2597	0.0007	1.1025–1.14392	—	—	—
Household's light source						
Gas and other sources	1.0000			1.0000		
Electricity	0.7523	0.0000	0.6691–0.8458	0.8394	0.0075	0.7382–0.9544
Household's water source						
Other water sources (well, etc.)	1.0000			—		
Pipe water	1.0089	0.8852	0.8947–1.1376	—	—	—
Household head's education						
Less than high school education	1.0000			—		
At least high school education	0.8289	0.0021	0.7354–0.9342	—	—	—
Presence of electricity in neighbourhood						
With no electricity	1.0000			—		
With electricity	0.8681	0.0324	0.7527–0.9882	—	—	—
Region						
Region 6	1.0540	0.4875	0.9086–1.2226	—	—	—
Region 7	0.9951	0.9440	0.8678–1.1411	—	—	—
Region 8	1.0000			—		

Source: Early Childhood Development Project Baseline Indicators Survey 2001 dataset.

Notes

a   Model 1 shows the results of the logistic regression for each covariate.

b   Model 2 shows the results of the parsimonious model.

c   Although marital and working status of mothers and type of residence are important variables in the analysis, the distribution of cases showed that nearly all the mothers are currently married, currently working and living in rural areas. Thus, the use of the marital status, work status and residence variables are no longer considered in the regression equation.

such as age, parity, education, household electricity and iron supplementation for previous pregnancy have been accounted for.

Discussion

Being a mover is arguably an important characteristic related to maternal anaemia. This study has shown that movers are healthier (not anaemic) than native-born mothers, supporting the healthy migrant argument that migration is a selective process favouring healthier populations. Although some studies (e.g. Teller 1973: 223) have implied that migrants who have assimilated into the community (place of destination) have similar access to health services, this study has shown that it is not the length of residence in the place of destination that is associated with anaemia, but rather characteristics attributed to being a mover or a non-mover. Migrants possess some 'unmeasurable' attributes which enable them to be resilient and adjust to a different environment, despite their length of residence in their place of destination.

The results also reveal that among the mothers, being native-born and young, having more than four children, being less educated, living in a household with no electricity and not taking iron supplements for prenatal care are attributes associated more with maternal anaemia. Younger mothers and less educated mothers tend to be anaemic since they have fewer concerns about their health. Having more children takes a toll on maternal nutrition and on nutrition stores, resulting in maternal wastage. Maternal nutrition is diminished, especially with more children. Thus higher parity is expected to be associated with maternal anaemia. The non-intake of iron supplements is an indicator of the mother's poor prenatal care practice and lack of awareness of the need for mineral supplementation during pregnancy. Therefore, in cases where mothers do not take iron supplements (even if these capsules are provided free by the public health system), they are less likely to seek prenatal care and, therefore, more disposed to health problems such as anaemia. The lack of electricity in the household suggests that household members may be more resigned to their current conditions, including their health. As Lia-Hoagberg *et al.* (1990: 493) argue, poverty's effect on seeking health care cannot be underestimated since poor people often feel that they have limited options and very little control over their life decisions, including their health. With all their problems, and with the need to earn a living and care for the household, they may find it difficult to give priority to prenatal care. This study has shown that even in disadvantaged areas, anaemia is a condition affecting those in poor households (without electricity), thus supporting the view that anaemia is associated with poverty.

The study's findings pin-point the need to support less educated, younger mothers; those with more children; those living in less favourable household conditions; those who do not take iron supplements; and those who do not seek prenatal care. Thus, intervention and information programmes should be geared towards these mothers in disadvantaged areas in order to curtail the maternal anaemia problem. Furthermore, the results imply that although one's migration status may play an important role in one's health status, living in a disadvantaged area should be a sufficient reason to provide equitable resources and health services. Both preventive and curative health programmes need to focus on the whole population, regardless of whether they are migrants or non migrants. The findings have indicated the presence of ill-health in the native-born population, showing that even without the migrants, the health service of the population is a major concern that needs to be addressed.

Acknowledgements

The author would like to acknowledge the Council for the Welfare of Children for the use of the dataset, Dr Socorro A. Gultiano for her magnanimous encouragement and support in this undertaking, and Director Josephine Avila and the staff of the Office of Population Studies for their generous support.

Notes

1 Those in need include populations with children aged 0–5 who are at risk of dying, or populations residing with children 6–12 years old who have dropped out of elementary school or who are underweight (less than 75 per cent of the standard). Those at risk include populations with children aged 0–5 who are living in households with limited information, in households with low per capita income, or in a community with limited social services (Council for the Welfare of Children 1999).
2 The selected *barangays* in the Program regions are those earmarked to be recipients of intervention programmes identified and supported by the local government, while those in the control region are not recipients of the same intervention programmes. Although there will be differences in the intervention programmes between the Program and control regions, these were still not in place during the Baseline Indicators Survey in 2001. Follow-up surveys in 2002 and 2003 and the End-line Indicators Survey (2004) are scheduled to capture the effect of these interventions.
3 According to the World Health Organization (1972), haemoglobin levels below 12.0 g/dl (for non-pregnant mothers and lactating mothers) and 11.0 g/dl (for pregnant women) are indicators of anaemia.
4 The cyanomethaemoglobin method was used to determine haemoglobin measurements from blood samples of mothers.
5 Only a few of the eligible sample mothers (0.14 per cent) refused to have their blood samples taken for haemoglobin-level analysis or were unable to have their blood samples taken.
6 To arrive at a parsimonious model, a stepwise regression was undertaken. Variables which showed a statistically significant contribution to the change in the deviation from the log likelihood statistic were included in the final model. The order in which the variables were included in the final model was based on their significant contribution ($p < 0.05$ level of significance) considering other covariates.
7 Other personal characteristics which are almost universal for both native-born and migrant mothers, and for anemic and normal mothers, were not included in

the analysis even if their relationship to anaemia may be important. These variables include marital status, work status and pregnancy status.

8 Although being in an urban or rural area could be an important covariate, it was excluded in the analysis because almost all the mothers were from rural areas.

References

Abejo, S. (1985) 'Migration to and from the National Capital Region, 1975–1980', *Philippine Journal of Statistics*, 36, 4: 7–22.

Abella, M. (1989) 'Policies and practice to promote migrants', *Philippine Labor Review*, 13, 1: 1–7.

Amacher, G., Cruz, W., Grebner, D. and Hyde, W. (1998) 'Environmental motivations for migration: population pressure, poverty and deforestation in the Philippines', *Land Economics*, 74, 1: 92–101.

Asis, M. (1990) 'Labor force experience of migrant women: Filipino and Korean women in transition', *Proceedings of The Expert Group Meeting on International Migration Policies and Studies of Female Migrants, San Miniato, Italy, 28–31 March 1990*, New York: United Nations.

Battistella, G. (1995) 'Data on international migration from the Philippines', *Asia Pacific Migration Journal*, 4, 4: 589–99.

Bollini, P. and Siem, H. (1995) 'No real progress towards equity: health of migrants and ethnic minorities on the eve of the Year 2000', *Social Science and Medicine*, 41, 6: 819–28.

Cabral, E., Hackenberg, R.A., Hackenberg, B.H., Magalit, H.F. and Guzman, S.V. (1983) 'Migration, modernization and hypertension: blood pressure levels in four Philippine communities', Paper presented at the 14th Annual Philippine Heart Association Convention, Cebu City, 25–28 May.

Chantavanich, S. and Paul, S. (1999) 'Reproductive health for migrant Burmese women in Ranong fishing community', *Development*, 42, 1: 73–4.

Cheong, R., Madriaga, J., Perlas, L., Desnacido, J., Marcos, J. and Cabrera, M. (2001) 'Prevalence of anemia among Filipinos', *Philippine Journal of Nutrition*, 42, 1–2: 45–57.

Council for the Welfare of Children (1999) *The Early Child Development Program Infokit*, Quezon City: Council for the Wefare of Children.

Dizon-Añonuevo, E. and Añonuevo, A. (2003) 'Women, migration and integration', Paper presented at the 25th Annual Scientific Meeting of the National Academy of Science and Technology, Manila, 9 July.

Duckett, M. (2000) 'Migrants and HIV/AIDS', *Development Bulletin*, 52: 18–20.

Eder, J.F. (1990) 'Deforestation and detribalization in the Philippines: the Palawan case', *Population and Environment*, 12, 2: 99–115.

Flieger, W., Koppin, B. and Lim, C. (1976) *Geographical Patterns of Internal Migration 1960–1970*, United Nations Family Planning Association – National Census and Statistics Office Monograph No. 5, Manila: National Census and Statistics Office.

Food and Nutrition Research Institute (2001) *Philippine Nutrition: Facts and Figures*, Tagig, MetroManila: Food and Nutrition Research Institute, Department of Science and Technology.

Go, E. (1995) 'Female migration in the Philippines', Paper presented at the 8th National Convention of Statistics, Manila, 4–5 December.

Gonzales, V.C. and Pernia, E. (1983) 'Pattern and determinants of interregional migration', in E. Pernia, C. Paderanga and V.P. Hermosos (eds) *The Spatial and Urban Dimension of Development in the Philippines,* Makati City: Philippine Institute of Development Studies.

Gupta, I. and Mitra, A. (1999) 'Knowledge of HIV/AIDS among migrants in Delhi slums', *Journal of Health and Population in Developing Countries,* 2, 1: 26–32.

Lia-Hoagberg, B., Rode, P., Skovholt, C., Orberg, C., Berg, C. Mullet, S. and Choi, T. (1990) 'Barriers and motivators to prenatal care among low income women', *Social Science and Medicine,* 30, 4: 487–95.

Littlefield, C. and Stout, C. (1987) 'A survey of Colorado's migrant farmworkers: access to health care', *International Migration Review,* 21, 3: 688–708.

Muennig, P. and Fahs, M. (2002) 'Health status and hospital utilization of recent immigrants to New York City', *Preventive Medicine,* 35, 3: 225–31.

Perez, A. (1978) 'Internal migration in the Philippines', *Population of the Philippines,* Country Monograph Series No. 5, Bangkok: UN Economic Social Commission for Asia and the Pacific, pp. 44–77.

Perez, A. (1985) *Migration in the Philippines, Past and Future,* Quezon City: University of the Philippines, Population Institute.

Pernia, E., Paderanga, C. and Hermosos, V.P. (eds) (1983) *The Spatial and Urban Dimension of Development in the Philippines,* Makati City: Philippine Institute of Development Studies.

Razum, O., Zeeb, H., Seval Akgün, H. and Yilmaz, S. (1998) 'Low overall mortality of Turkish residents in Germany persists and extends to second generation: merely a healthy migrant effect?', *Tropical Medicine and International Health,* 3, 4: 297–303.

Rush, D. (2000) 'Nutrition and mortality in the developing world', *American Journal of Clinical Nutrition,* 72, 1 (Supplement): 212S–240S.

Sastry, N., Goldman, N. and Moreno. L. (1993) 'The relationship between place of residence and child survival in Brazil', *International Union for the Scientific Study of Population, International Population Conference, Montreal,* 24 August to 1 September: 293–321.

Scholl, T. and Reilly, T. (2000) 'Anemia, iron and pregnancy outcome', *Journal of Nutrition,* 130 (Supplement): 443S–447S.

Sembrano, M.A. (1980) *Migration, Fertility and Social Mobility in Rural and Urban Communities,* Southeast Asia Population Research Awards Program (SEAPRAP) Report No. 63, Singapore: Institute of Southeast Asian Studies.

Sims, J. (1994) *Anthology on Women, Health and Environment,* Geneva: World Health Organization.

Steer, P. (2000) 'Maternal hemoglobin concentration and birth weight', *American Journal of Clinical Nutrition,* 71 (Supplement): 12855S–12875S.

Stoltzfus, R.J. (2001) 'Iron deficiency anemia: re-examining the nature and magnitude of the public health problem – Summary: implications for research and programs', *Journal of Nutrition,* 131 (Supplement): 697S–700S.

Teller, C. (1973) 'Access to medical care of migrants in a Honduran city', *Journal of Health and Social Behavior,* 14: 214–326.

World Health Organization (1972) *Nutritional Anemia,* Technical Report Series No. 503, Geneva: World Health Organization.

World Health Organization (2002) *The World Health Report 2002: Reducing Risks, Promoting Healthy Life,* Geneva: World Health Organization.

7 The Filipinos in Sabah

Unauthorized, unwanted and unprotected

Maruja M.B. Asis

Introduction

The massive deportations carried out by Malaysia in 2002 called attention to the manifold issues and tangled realities of unauthorized migration.[1] Alarmed by the unabated presence of unauthorized migrants – often referred to as 'illegal migrants' or 'illegals' – Malaysia resolved to amend its Immigration Act to deal decisively with unauthorized migration by way of more punitive measures.[2] Effective from 1 August 2002, the amendments provide for unauthorized workers to be caned, gaoled for six months and/or fined M$10,000. The riots staged by Indonesian workers in January 2002 provided the Malaysian government with a rationale to hasten the crackdown, beginning with the decision to make Indonesians a last priority in the hiring of migrant workers. Indonesians comprise the majority of some 1.2 to 1.7 million migrant workers in Malaysia.[3] Discourses on unauthorized migration as a 'problem' were heard in Malaysia, as well as in the Philippines and Indonesia, the countries most affected by the forced repatriations of their nationals. In Malaysia, the problem was linked to enforcement and national security, concerns which justified the deportations. Meanwhile, in the Philippines and Indonesia, media reports chronicled the tragic toll of the deportations, especially as the deadline approached. Malaysia's unrelenting stance strained relations with its neighbours. In the Philippines, outrage over the treatment of Filipinos reignited calls to review its claim to Sabah.[4]

Closer to the deadline, the arrests and deportations intensified, resulting in overcrowding in detention centres. Migrants rushed home, congesting ports, ships and reception areas. Malaysia eventually extended the grace period to 31 August. In the rush to beat the deadline, Nunukan Island in East Kalimantan became emblematic of the humanitarian crisis created by the new policy. An exit and entry point for Sabah, Nunukan found itself swamped by some 20,000 to 30,000 (another report cited 40,000) Indonesians expelled from Malaysia. As thousands of migrants descended on Nunukan, reports detailed the health conditions of migrants. Makeshift camps were set up – many more migrants slept in whatever open spaces they could find, where lack of water and unsanitary conditions made them susceptible to health problems. As of 31 August, as many as sixty-three migrants, including children, had died. The major causes of death were meningitis, malaria, diarrhoea, typhoid and respiratory diseases (*Asian Migration News*, 15 September 2002).

In the Philippines, migrants arrived either in Zamboanga City or Tawi-Tawi. The local government in these areas had for many years routinely dealt with deportations. However, the 2002 deportations were different because they attracted national attention. In addition, non-government organizations (NGOs) were visibly involved in responding to the emergency and rehabilitation needs of the deportees. At the height of the deportations, government agencies and NGOs carried out several funding drives and initiated reintegration programmes for the returnees, including *Tabang Mindanaw* (Help Mindanao), and *Operation: Pagbabalik* (Operation: Homecoming). Following reports of overcrowding in detention centres, lack of food and water during the deportation process, and deportees getting sick or dying, the Philippine government filed a formal protest with Malaysia. The latter acceded to the Philippine government's request to send a delegation to investigate the conditions in Sabah's detention centres.

A year later, reports on crackdowns, arrests, detentions and deportations continued, but they were no longer attention-grabbing. Such come-and-go attention given to the issue of unauthorized migration seems to be the norm. Unauthorized migration is tolerated, kept under wraps or ignored for the most part, until some crisis – whether real or imagined – emerges and unauthorized migrants are targeted. As sources of information on unauthorized migration, highly publicized deportations highlight unusual conditions and, as such, the information that the media reports may not reflect the day-to-day experiences of migrants.

Focusing on the Filipinos in Sabah, I examine the experiences of unauthorized migrants under relatively ordinary circumstances. How do unauthorized migrants live and work in the country of destination outside the glare of publicity? In particular, how does their unauthorized status affect their health and access to health care at different stages of migration; that is, from the time of their exit from the Philippines to their entry and stay in Sabah, and to their return to the Philippines?

Data from three research projects carried out by the Scalabrini Migration Center provide some insights into the health conditions, health hazards and access to health care of unauthorized Filipino migrants in Sabah. The data sources include, first, an exploratory study which mapped the Western Mindanao-Sabah migration system (the '1997 study').[5] This study was part of a larger study on migration from the Philippines to Malaysia and involved fieldwork in Zamboanga City, the staging point of legal migration from Mindanao to Sabah, and Sitangkay (Tawi-Tawi), one of the staging points of undocumented migration to Sabah. Interviews were conducted with traders and visitors based in the Western Mindanao area, deportees (*balaws*) and returnees (that is, those who returned voluntarily to the Philippines).

A second data source for this research is a study of HIV vulnerability in the Western Mindanao-Sabah migration system (the '2000a study'). This study examined the factors contributing to HIV vulnerability, as well as resilience. Fieldwork was carried out in Zamboanga City, Bongao (Tawi-Tawi), a key exit/entry point, and Kota Kinabalu and Sandakan (Sabah, Malaysia). In addition to interviews with former migrants to Sabah, interviews were also conducted with Filipinos based in Sabah – legal long-term settlers, refugees and professionals, and unauthorized domestic workers, musicians and professionals. Observations and informal interviews were also conducted with Filipinos who were working in hotels, restaurants, karaoke bars and markets. During fieldwork in Sabah, migration-related articles published in English newspapers were monitored.

The final data source for this research is a study of unauthorized migration from the Philippines (the '2000b study'). This was part of a four-country study on unauthorized migration in Southeast Asia (the three other participating countries were Indonesia, Malaysia and Thailand). Data collection in the Philippines was conducted in the second half of 2000 (Battistella and Asis 2003b). The study purposely did not include the Mindanao-Sabah component because of its unique character (as detailed in the 1997 study). However, during fieldwork in Capiz, a province in the Visayas which was one of the research sites, the researchers discovered a group of migrants who had been to Sabah. These migrants were interviewed because their migration to Sabah was outside traditional cross-border flows involving the people of Western Mindanao.

Since these studies were not designed specifically to examine the relationship between migration and health (except for the 2000 study on migration and HIV/AIDS concerns), the collection of data was not focused on this question. Thus, the data generated from these studies are largely qualitative, which limits the analysis to an exploratory one. The first part of this chapter frames the different health outcomes implied by legal versus unauthorized migrations. Specifically, the discussion explores how the legal framework of migration provides some basic guarantees in promoting the health and well-being of migrants. This is followed by an overview of the historical and contemporary contexts of the Western Mindanao-Sabah migration system. In the third section I focus on the working and living conditions of unauthorized Filipino migrants in Sabah, and on how their legal status may or may not affect their health-related conditions.

Migration, health and legal status: some considerations

The legal status of migrants is a factor that can mediate the relationship between migration and health. In legal migration, migrants are subject to health screening and surveillance. Although legal migration imposes requirements and exacts costs on migrants, it also provides them with some basic guarantees. In contrast, the lack of regulation in unauthorized migration predisposes migrants to conditions that do not promote safety, good health and access to support in the case of health problems.

Under legal labour migration, prospective overseas Filipino workers have to go through three key checks: pre-deployment health checks, travel checks, and health checks and monitoring in the country of destination, each of which is dealt with below.

The common perception of migrants as harbingers of disease and contamination departs from the notion of migration as a selective process. Migrants are supposedly healthy; otherwise they are screened out (hence, the 'healthy migrant effect'). Legal migration requires that prospective migrants pass a medical examination as a requirement for admission to the country of destination. Particularly for migrant workers, the medical examination determines whether they are fit for employment. In the Philippines, for example, applicants for overseas employment must submit to a medical examination after their placement agency has found an employer for them, the results of which are valid for six months.[6] In cases where an applicant cannot be deployed within the said period, he or she has to undergo further screening.

The basic medical examination involves a complete physical examination; a chest X-ray (valid for three months); a complete blood count, including haemoglobin determination; blood typing; urinalysis; a stool examination; a dental examination; a psychological evaluation; and a blood chemistry analysis. Some countries or some job categories may require additional tests. In the case of nurse applicants, for example, they must also undergo HIV/AIDS-antibody testing,[7] Hepatitis A and B, visual acuity and pregnancy tests.[8] Female applicants may be required to retake a pregnancy test should there be a delay in their deployment. The package of tests costs at least P2,700, which applicants pay to the agencies.[9]

A medical clinic or hospital is accountable for the medical certificate that it issues, a precautionary measure against anomalous practices. According to AO No. 85–A, Revised Rules and Regulations Governing Accreditation of Medical Clinics and Hospitals and the Conduct of the Medical Examination for Overseas Workers and Seafarers (Department of Health 1999: 7–8),

> The hospital or medical clinic shall guarantee the integrity of its medical examination results for a period of three months. . . . The cost of repatriation and other expenses shall be shouldered by the clinic in the event the employer rejects the accepted applicant or terminates the employment of the workers/seafarers within a period of three (3) months from the date of examination due to medical reasons.[10]

In addition to the medical examination, a legal migrant worker is also protected by a contract, which among other things specifies minimum working and living conditions and the employer's responsibility to provide the worker with insurance.[11] The migrant and his or her family are also covered by medical insurance upon the migrant's enrolment in the medicare programme.[12] Prior to their departure, legal migrant workers are required to attend a pre-departure orientation seminar which provides them with information about the country in which they will work, working and living conditions, services and assistance that they can access, and health information.[13] Unauthorized migrant workers are free of these requirements and regulations. However, they are rendered vulnerable in many ways: leaving without knowing whether they are fit for employment; not having a contract to guarantee basic minimum working and living conditions; not having health coverage; and lacking information about the conditions of working abroad and access to support.

Legal migrant workers are generally assured of safe travel conditions; unauthorized migrant workers have no such guarantee. Those who had gone to or returned from Sabah via the back door spoke of the physical dangers and anxieties they experienced in crossing the border. Transport overcrowding, encountering pirates, being exposed to the elements, evading coast guard authorities, or having to wade through waters in the dark to avoid the authorities were part of their ordeal. In the event that they come to harm, no one is held accountable.

Some countries of destination, such as Malaysia, require another medical examination when migrant workers arrive, which may be followed by periodic checks during a worker's stay in the country. Since 1997, Malaysia has centralized the monitoring and supervision of migrant workers' medical examinations through the Foreign Workers Medical Monitoring Agency, a system introduced in Sabah in 1998. Like Singapore, Malaysia requires female migrants to undergo a pregnancy test. Workers found unfit or pregnant are repatriated to their countries of origin.

Upon finishing their contract, workers are supposed to return to their countries of origin. The return phase does not require health monitoring, either by the country in which they have worked or their home country. Those who are repatriated for health-related reasons assume responsibility for any medical intervention they need. Those who return at the end of their contracts are generally no longer covered by health insurance. It is during this phase of migration when legal and unauthorized workers find themselves in a similar situation; that is, they are basically on their own should health problems arise.

Thus, the processes involved in legal versus unauthorized migration set some preconditions on the possible health trajectories of legal and unauthorized migrants. Falling outside the framework of legal migration, unauthorized migrants and travellers using the back door are unencumbered by the health surveillance and monitoring of the state. The system can be a form of control and adds to the costs of migration, but at the same time it provides a means to detect and screen out those who are not fit to work and it provides some measure of support for migrants when they fall ill.

Studies that have examined the health conditions of unauthorized migrants typically show that unauthorized migrants are indeed vulnerable to health problems, exacerbated by lack of access to health services. Poor working and living conditions and lack of access to health care (that is, lack of health insurance) described the conditions of unauthorized Indian workers in Lebanon (Gaur and Saxena 2004). Migrants who fell ill did not seek medical help not only because they lacked insurance, but also because they feared they would be deported (Gaur and Saxena 2004). In the case of Myanmar migrants in Thailand, while they have access to Thai health care services, they were less likely than Thai citizens to use government facilities (Isarabhakdi 2004). Legal status, financial costs, location of facilities and their operation hours and inability to speak Thai are factors that prevent migrants from availing themselves of government services. They do have access to these services but only if they pay. Aside from logistics, belief in spirits determines the health-seeking practices of Myanmar migrants. The Karens, who believe that illnesses are caused by spirits, are more likely to resort to traditional healers than are Mon and Burmese migrants (Isarabhakdi 2004).

The vulnerability of unauthorized migrants to health problems is not unexpected, but how they deal with health issues given their unauthorized status needs to be explored further. In the case of Filipinos in Sabah, their long history of migration and *de facto* settlement raises questions about alternative courses of action in solving their health problems: What do they do when they have health problems? What is the role of ethnic networks, if any, in reducing their health vulnerabilities? What are the other factors that lessen their vulnerabilities to health risks?[14]

From traditional movements to unauthorized migration

It is important to situate Filipino migration to Sabah as an extension of traditional movements, and as part of other migrations taking place in Southeast Asia. A major component of contemporary migrations is the movement of workers to the more vibrant economies in the region and beyond. Three migration subsystems have been established since large-scale labour migration commenced in the 1970s (Battistella and Asis 2003a) (Figure 7.1).

First, in mainland Southeast Asia, Thailand is the core country attracting migrants from neighbouring Burma (approximately 80 per cent), Cambodia and Laos. Most of the migrants in Thailand are unauthorized migrant workers. Second, on the Malay Peninsula, Singapore and Malaysia are the core countries and both countries have a substantial share of foreigners in the workforce – approximately 29 per cent in Singapore, and 16 per cent in Malaysia. Singapore's migrant workers come from the Philippines, Indonesia, Thailand and South Asia on a contract system. In addition, thousands of Malaysians cross the border daily to work in Singapore's industries, an arrangement that has been in place since 1978. Malaysia established a legal channel for labour migration in 1991; however, large numbers of

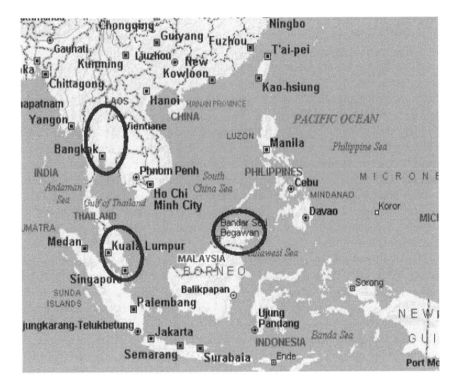

Figure 7.1 Migration systems in Southeast Asia (source: Battistella and Asis (2003a: 7)).

Indonesians and Filipinos had already arrived in Malaysia before the system was introduced. Later, many more migrants arrived through unauthorized channels. Consequently, there are more unauthorized migrants in Malaysia than legal ones. Third, in maritime Southeast Asia, Brunei and Sabah (East Malaysia) are core areas. While migration to Brunei is regulated by a work contract system, in Sabah it is largely unauthorized. Brunei's migrant workers come from other Association of South East Asian Nations (ASEAN) countries (including Malaysia), while Sabah's come from the Philippines and Indonesia. A subregional cooperation, the BIMP–EAGA (Brunei-Indonesia, Malaysia, Philippines–East ASEAN Growth Area), was formed in this part of Southeast Asia in 1994 to promote development in the region and has led to increased migration.[15]

Two patterns are discernible in the migration subsystems in Southeast Asia: most migrations occur in the region (notably, most are cross-border flows between adjacent areas), and unauthorized migration is significant. History, geography and culture have some bearing on these patterns. To begin with, the history of Southeast Asia is replete with migration themes. According to Wright (1990: 43):

General histories about the region create an impression of great populations constantly on the move from one place to another. They focus on the greater and lesser migrations and conquests, the endless ebb and flow of refugees, nomads, vagabonds, and sea-gypsies that constitute the major patterns of migration in Southeast Asia.[16]

Prior to the arrival of Western powers, maritime Southeast Asia was a seamless Malay world or *Dunia Melayu* wherein trade and flows of peoples were commonplace. Unlike the north-to-southward movement of people in mainland Southeast Asia, the migrations of peoples in maritime Southeast Asia were multi-directional. In this part of Southeast Asia, Mindanao and Sabah were sites of intense and varied forms of population mobility, including piracy (Wright 1990; Warren 2003). Wright (1990: 45) notes: 'There have been traditionally Malay peoples of a nomadic bent throughout the region encompassing the southern Philippines and northern Borneo.'

These fluid exchanges developed relatively stable linkages. Today, the seamless world of maritime Southeast Asia has been cut into different parts belonging to Indonesia, Malaysia and the Philippines – and the hitherto unfettered movements of people have been subjected to the rules of international migration. Population mobility continues, but most of it is not in keeping with legal international migration. The high levels of unauthorized migration in this subregion (and throughout Southeast Asia), most of which are cross-border flows, are indicative of the persisting logic of history, geography and shared culture in the border areas.[17] Wong and Teuku Afrizal (2003) proposed that these cross-border flows may be more aptly viewed as regional migration, rather than international migration. As regional migration, the movement of people should not be regulated according to the system of international migration, a scenario that could open up new and more lasting solutions to the problem of unauthorized migration.[18]

Although history, geography and cultural affinity form an important backdrop to migration from Mindanao to Sabah, present-day migrations to Sabah also have economic roots. The perception of job opportunities and better incomes in Sabah is an important factor, drawing not just the peoples of Western Mindanao to Sabah, but also Filipinos from other regions in The Philippines. As is shown below, the peace and order situation in Mindanao is another factor that compels the flight of people to Sabah. I return to these points in the next section.

The Western Mindanao-Sabah migration system

The Western Mindanao-Sabah migration system is popularly known in the Philippines as the 'southern back door'. Fact, fiction, legend and lore have associated this migration trail with shadowy activities and figures – piracy, smuggling, fugitives escaping from the reach of the law, and the entry point of foreigners entering the Philippines without travel documents.

Far from 'imperial Manila', the comings and goings of people via the south-ern back door usually go unnoticed.[19]

From the Sulu archipelago, specifically from the province of Tawi-Tawi (Bongao, Sitangkay and Taganak Island are some of the staging points), the centuries-old path to Sabah continues to be plied by the Tausugs, Samals, Badjaos and Yakans to visit family and friends, to trade or to find work in Sabah (see Figure 7.2).[20] These migrations are extensions of the traditional migrations trod by their ancestors. Interviews with residents in the area revealed that they did not view Sabah as a 'foreign' place. Respondents referred to travels to Sabah as 'our way', and going to Sabah as a practice that had existed 'since time immemorial'. Particularly for those who live along the border, Sabah is more familiar than foreign. The residents of Taganak Island, located at the edge of the international treaty limits separating Malaysia and The Philippines, for example, have more links to Malaysia (Sandakan, Sabah) than to The Philippines. They market their produce in Sandakan, their stores are stocked with Malaysian goods, and the preferred currency is the Malaysian ringgit (http://oneocean.org 2003).[21] Although Bongao and Sitangkay are not as close to Sabah, the stores in these areas display staples and goods from Malaysia.

Contemporary migrations to Sabah, however, are not just a continuation of old patterns. Economic factors underlie more recent migrations. Min-danao, the traditional region of origin of migrants to Sabah, was neglected by the national government until recently. Although some parts of Min-danao have achieved remarkable economic growth, the rest lags behind. Western Mindanao provinces, in particular, have been among the least economically developed areas in the Philippines. As Table 7.1 shows, Basilan, Tawi-Tawi and Sulu rank lowest among the seventy-seven provinces in the Philippines in terms of selected development indicators. Families, therefore, have used migration to Sabah as a strategy to augment their income. The presence of family and kin in Sabah eases the passage to Sabah and into the labour market. Community leaders and officials interviewed in Tawi-Tawi in the 1997 and 2000a studies opined that people would con-tinue to cross the border. They were pessimistic about the effectiveness of deportations and repatriations, since those who were deported planned to find a way to return to Sabah. Thus, instead of deportations, they hoped that cross-border agreements between the Philippines and Malaysia would be worked out, an arrangement that would promote safer passage.

Job opportunities and better wages in Sabah have also lured less tradi-tional migrants, that is, residents from outside Western Mindanao, to try their luck in Sabah. Middlepersons and brokers have advertised job opportunities and attractive wages in Sabah and arranged for passage – for a fee. Since residents of Western Mindanao are generally familiar with the route, brokers are most likely to entice those outside Western Mindanao. Brokers may arrange passage via *kumpit*,[22] in which case travel/work docu-ments are not necessary, or they may tap the ferry service – the legal way to

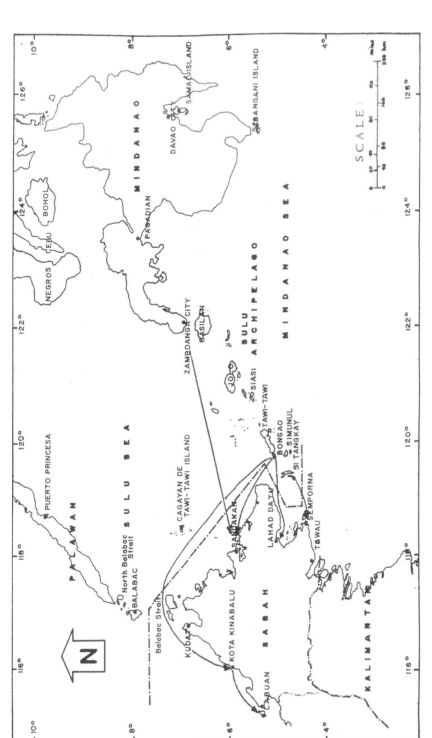

Figure 7.2 Pathways from West Mindanao to Sabah (source: Battistella *et al.* (1997: 62)).

Table 7.1 Western Mindanao provinces: selected indicators

Measure/indicator	Value	Rank[a]
Human Development Index, 2000		
Basilan	0.420	75th
Tawi-Tawi	0.378	76th
Sulu	0.311	77th
Per capital income (NCR 1997 Pesos), 2000		
Basilan	13,193	74th
Tawi-Tawi	11,349	76th
Sulu	7,850	77th
Gender-related Development Index, 2000		
Basilan	0.406	74th
Tawi-Tawi	0.366	76th
Sulu	0.322	77th
Poverty incidence, 2000 (Phils = 27.5)		
Basilan	63.0	73rd
Tawi-Tawi	75.3	76th
Sulu	92.0	77th
Life expectancy at birth, 2000		
Basilan	60.2 yrs	75th
Tawi-Tawi	50.8 yrs	76th
Sulu	52.3 yrs	77th

Source: *Human Development Network and UNDP* (2002).

Note
a Rankings for seventy-seven provinces.

leave the Philippines and to enter Sabah. As ASEAN members, Filipinos travelling to Sabah do not need a visa to enter Malaysia. Social visit pass holders violate the terms of their admission into Sabah once they work and/or overstay. Other social visit pass holders get around the restrictions on their stay by exiting Sabah – mostly to Brunei – before their social visit pass expires, and re-enter Sabah with a new 'chop' or a new permit of stay.[23]

The opening of the ferry service between Zamboanga City and Sandakan (Sabah) in 1995 is one of the fruits of the BIMP–EAGA. By 2000 there were two shipping companies providing ferry services. In part, the BIMP–EAGA was also an attempt to promote the movement of people through regular channels.[24] Regular ferry services, however, have not significantly reduced the passenger traffic via *kumpit*. During fieldwork in Bongao in 2000, the schedule of the trips to Sandakan was widely known in the community. Similarly, in Sandakan, there was open advertisement of the schedule of departures for Bongao.[25]

Before the 1970s, Filipinos were received positively in Sabah. In the

1960s, Datu Tun Mustapha, who served as Chief Minister (1967–1976), encouraged Filipino Muslims to settle in Sabah (Battistella *et al.* 1997). In addition, Sabah had a tradition of recruiting Filipinos before they became global workers. Logging and construction companies hired Filipinos, some of whom came with, or were later joined by, family members. This cohort of Filipino workers and their descendants are among the earliest documented Filipinos in Sabah. A descendant whom I met in Sabah in 2000 said proudly: 'Filipinos helped build Sabah.'

In the 1970s the arrival of Mindanao refugees was a turning point in the social standing of Filipinos in Sabah. Fleeing from conflict, triggered by the burning of Jolo in 1974, the arrival of thousands of Filipino Muslims in the 1970s increased the Filipino population in the state.[26] The large influx of refugees and the possibility that they would stay permanently in Sabah raised concerns about the state's ethnic balance. Later, this was reinforced by the arrival of 'refugees' of another kind – Filipinos who came looking for work. Where before the authorities generally tolerated the presence of Filipinos, by the 1980s Sabah had started to take a more vigilant stance in guarding its borders. Since then, the government has implemented crackdowns, deportations, border controls and regularization to deal with unauthorized migration.

Sabah's changing response to Filipinos reflected growing unease in Malaysia about the presence of foreigners, specifically the 'illegals', and their impact on society. Numbers can affect the ethnic balance in a multi-ethnic society such as Malaysia. Sabah's population – which includes indigenous populations, of which the largest group is the Kadazan/Duzun – grew increasingly worried. The Parti Bersatu Sabah (PBS) was very vocal in airing concerns about the influx of unauthorized migrants.[27] The perception by some Sabahans that the United Malays National Organization was encouraging the entry of Filipinos, specifically Filipino Muslims, to expand its political base, also created a backlash against Filipinos.[28] The alleged participation of 'illegals' in state elections was a contentious point. Some Sabahans, including some long-term Filipino settlers interviewed in 2000, expressed alarm that non-Sabahans were determining the political future of Sabah. In addition, there were the usual anxieties attributed to migrants – they take jobs away from local workers; they overburden the state's social services; they are responsible for the rise in criminality. The media in Sabah generally depict Filipinos in a negative light. Filipinos engaging in crimes, drug dealing, squatting or begging are the usual fare (see *Asian Migration News* various issues; Kurus 2002).[29]

Sabah's resource-rich economy, however, requires workers, a need which has historically been met by migrant workers.[30] According to the State Labor Department, migrant workers comprised 55.8 per cent of the total labour force in Sabah in 1996, up from 35 per cent, 52.2 per cent and 58.3 per cent in 1970, 1980 and 1990, respectively (Kurus 2002: 70). Sabah's problem is twofold: it has a high unemployment rate, but at the same time it lacks workers in the palm oil plantation, construction and domestic work sectors. In other words, unemployment is caused not so much by lack of jobs as by local workers preferring to work in other sectors. Employers have thus turned to migrants to perform jobs in the sectors shunned by local workers. During fieldwork in Kota Kinabalu and Sandakan, Filipinos were indeed ubiquitous in Sabah, particularly in the service sector. At the start of the 2002 deportations, Datuk Thomas Lau Chi Keong, former Deputy President of the Sandakan Municipal Council, acknowledged that Filipino workers were needed in Sandakan and Kota Kinabalu – but he added that they must be legal workers (*Asian Migration News* 15 April 2002).

Figures on the size of the undocumented population in Sabah in general, and the Filipino population in the state in particular, are difficult to pinpoint. Based on the 2002 census, non-citizens (estimated at 614,824) account for some 25 per cent of Sabah's total population of 2,603,485 (Kurus 2002: 69). Sabah's non-citizen population accounts for 44.4 per cent – the largest share – of Malaysia's non-citizen population of 1,384,744 (Kurus 2002: 69). Data from the Federal Special Task Force regularization programme in 1997 estimated 585,796 foreigners in Sabah, including Indonesian and Filipino migrant workers and their dependants who had registered, Indonesian migrant workers who had work permits, refugees (most of whom were Filipinos who arrived in the 1970s), those who failed to register, and other foreign nationals (Kurus 2002; 1999). In 1997, *at least* 216,000 Filipinos were in Sabah (see Table 7.2). Estimates from other sources suggest higher figures. Interviews with key informants in the Philippines yielded estimates ranging from 500,000 to 1 million (Battistella *et al.* 1997). A 2001 government estimate of the irregular Filipino population in

Table 7.2 Estimated foreign nationals in Sabah, 1997

Category	Number
Migrant workers and their dependants	413,832
(Filipino migrant workers and their dependants)	(119,128)
Existing stock of Indonesian workers with work permit (before the regularization programme)	30,767
Refugees	57,197[a]
Estimated undocumented foreign nationals (that is, those who failed to register)	80,000
Other foreign nationals (Indians, Pakistanis)	4,000
Total	585,796[b]

Source: Table 1 in Kurus (1999:283)

Notes
a Mostly Filipinos who arrived in the 1970s.
b Assuming that Filipinos comprised at least half of the estimated foreign nationals and all of the refugees, at least some 216,000 Filipinos were in Sabah as of 1997.

Malaysia was 509,524 (Battistella and Asis 2003a: 40). Since migration to Peninsular Malaysia is mostly legal, this estimate would refer to Filipinos in Sabah. The estimates of unauthorized migrants in Sabah ranged from between 150,000 and 500,000 (see *Asian Migration News* 15 April 2002).[31]

Due to the large numbers of unauthorized Filipinos in Sabah, 'illegal' has come to be regarded as the typical Filipino in Sabah. This stereotype masks the diversity of the Filipino migrant population. This diversity and the lack of interaction among the different groups suggests that there is no one Filipino community in Sabah. Rather, there are various Filipino communities which are identifiable along the lines of arrival in Sabah, ethnic background in the Philippines, socio-economic status and so forth. I was also quite struck by the way in which Filipinos in Sabah identified or referred to co-nationals not as Filipinos, but in terms of their ethnic membership. Table 7.3 summarizes the key variables describing the Filipino population in Sabah. The most crucial variable – and the one identified by all respondents as the primary problem of Filipinos in Sabah – is their legal status.

The undocumented life: daily realities

For the majority of unauthorized Filipino migrants in Sabah, their daily life is plagued by many uncertainties. According to respondents, inspections at checkpoints and raids are commonplace. Raids take place any time, anywhere. Interviewees who had been deported to Zamboanga City and Tawi-Tawi claimed they had been randomly apprehended by police and deported to the Philippines. Some had been repatriated without their family members' knowledge. Thus, when asked about the problems they faced in Sabah, the singular most common answer was their undocumented status. Being undocumented pervades various aspects of their life – receiving sub-standard wages, lack of job security, not having the liberty to return to the Philippines on a more regular basis, as well as access to health care. Many also expressed the view that Sabahans 'looked down' on Filipinos.

Despite the uncertainties and insecurities of the unauthorized life, respondents claimed that jobs were not difficult to find in Sabah if they were not choosy. Several respondents' work histories suggested that they started working as children. Many respondents had managed to stay in Sabah for a relatively long period of time. Some interviewees in Kota Kinabalu had been working there for many years: Miguel, a waiter, arrived in 1978; Susan, a restaurant worker, arrived in 1990, and had been joined by a brother; Ning, also a restaurant worker, had come to Sabah in the 1990s.[32] Susan related encounters with police and considered herself fortunate that she had been able to pass herself off as Malay. Ning claimed that her mother bought an identity card for her.[33] Another strategy of getting around documentation problems was suggested by Bunny's experience. A singer in Sandakan since 1994, Bunny had arrived on a social visit pass. She said that she would go to Brunei before her permit expired, stay there for a day, and re-enter Sabah

Table 7.3 Filipinos in Sabah: a classification scheme

Criterion	Subgroups	Remarks
Legal status	Documented permanent settlers and residents	—
	Those who were recognized as refugees	—
	Long-term settlers and residents without documents	Includes traditional migrants who did not secure documents during their residence in Sabah
	Migrants with work permits	As of 2002, there were 10,122 legal foreign workers (Azizah (2002), cited in Kurus, 2002: 70)
	Undocumented migrants (no travel/work papers)	—
	Overstayers/tourists-turned-workers (migrants who initially had travel papers but who overstayed and worked in the state)	—
Region of origin in the Philippines (a proxy for ethnicity)	Traditional migrants: Tausugs, Samals, Badjaus, Yakans from Western Mindanao	Movements could range from short-term visits to more long-term stays in Sabah
	Migrants from other regions in Mindanao	May include other Filipino Muslims from other parts of Mindanao
	Migrants from other regions in the Philippines (Luzon and the Visayas)	—
Year of arrival	Before the 1970s	—
	Early 1970s	Arrival of refugees from Mindanao
	Late 1970s and after	Includes also trafficked persons
Occupation (a proxy for socio-economic status)	Professionals	—
	Service workers (including domestic workers, restaurant workers)	—
	Construction workers	—
	Entertainment workers, including band members	Includes trafficked persons
	Plantation workers	—

with a new 'chop'. Alternatively she would return to Zamboanga City and re-enter Sabah and get a new chop. She claimed she obtained new chops or extensions by paying off border personnel (Interviews, Sandakan, 2000a study).

For the many who are able to evade or escape the authorities, there are as many who are not as lucky. Ning's brother was picked up by the police and deported in 1999. Her brother had been writing to ask her to send him money so that he could return to Kota Kinabalu (Interview, Kota Kinabalu, 2000a study). A Badjau construction worker, interviewed in Tawi-Tawi, said that he had been in Sandakan for about seven years before he was deported in 1999. He had been detained by the police on suspicion of theft, during which time he was interrogated and whipped. When he complained about the whipping, the police officer responsible was dismissed. Although the construction worker was eventually cleared, he was deported. His wife and child later followed him to Tawi-Tawi, but they planned on returning to Sandakan as soon as possible.

While jobs are not difficult to find, decent wages and working conditions are more elusive. Not surprisingly, most migrants can ill afford to find a healthy environment in which to live. Construction workers usually live on the construction site; likewise plantation workers are usually provided with bunkhouses; other migrant workers who are not provided with housing by their employers rent rooms in crowded *kampongs* (villages). Migrants' settlements are also notorious for their unsanitary conditions. The following observation is routinely reported in the media. In a call to take drastic action against the presence of 'illegal migrants', Sembulan Assemblyman Edward Yong Ooi Fah was reported to 'attribute the negative attitude of foreigners and illegal immigrants to the massive accumulation of garbage in their dwelling areas, such as Sembulan, Putatan and Pulau Gaya' (*Daily Express* 24 April 2003). In addition to deplorable conditions, migrant communities are also reputed to be high crime areas. In Sandakan, for example, BDC, a community populated by Filipinos, is reputedly 'crime-infested'.

The low wages received by unauthorized migrant workers impose constraints on securing decent housing and access to other basic necessities. The examples of Susan and Ning illustrate the difficulties of making ends meet (Interviews, Kota Kinabalu, 2000a study). Although both worked in restaurants, their wages and work hours were different. According to Susan, she earned RM500 a month working from 4 p.m. to 10 p.m. Out of her monthly earnings she allocated RM150 for rent (a room). The rest went on food, other living expenses and remittances. Ning, on the other hand, earned RM250 a month, working from 8.30 a.m. to 8.30 p.m. daily, with two days off each month. Of her monthly income, she paid RM100 rent and RM50 for transportation. Ning lived with her mother and stepfather, and she described her life in Sabah as 'pitiful'. She expressed a desire to study, but she could hardly save from her earnings.

The most recent opportunity to regularize unauthorized migrants' status came in 1997. Susan and some domestic workers interviewed in Kota Kina-balu availed themselves of the 1997 regularization programme (Interviews, 2000a study). Regularizing their status gave them the freedom to move about without fear of apprehension. On the other hand, there were also disadvantages, such as having to pay for the medical examination (RM190) and the levy (RM460), which were deducted from their salaries. Another disadvantage of being legal was the fact that they could not change employers, even if their work conditions were not satisfactory. Thus, when their permits lapsed two years later, they continued to stay on and work without proper documents.

Due to low wages, migrant children also start work at an early age. Aside from the family's need to pool resources, migrant children may be driven to seek work because they cannot attend state-funded or supported schools. In 2002, the state required children to present education and immigration permits before they could be admitted to government schools. Former Chief Minister Harris Saleh expressed the view that the new policy would deprive 30,000 Filipino-Muslim children of schooling. He recommended that children born in Sabah, and those who had been attending schools for some time, should be exempt from the ruling (*Asian Migration News* 31 January 2002, 15 February 2002). In the course of the three studies, I encountered several migrants who used to be child workers in Sabah. The Badjau deportee, mentioned earlier, went to Sabah when he was about eleven years old. He used to help his mother (a vendor) and later joined his uncle in construction (Interview, Tawi-Tawi, 2000a study). Anna, a returnee interviewed in Tawi-Tawi, said that she and her siblings worked in Tawau. She and her brother worked in a grocery store – she earned RM160 while her brother earned RM100 a month. Her two other sisters worked in a *kedai* (store) and their combined income was RM270. She and her siblings returned to Tawi-Tawi in 1998 so that they could study. Anna described their life in Sabah as 'difficult'. She said that she was happier in Tawi-Tawi because she and her siblings could go to school.

Aside from low wages, unauthorized migrants are also vulnerable to abuse. One of the respondents in the 2000b study, a former domestic worker, said that her employers subjected her to very long working hours, did not provide her with food and did not pay her. She ran away after three months and lived with a Filipino family until she received money from her family to buy a return ticket. Another case involved twenty-three migrants from Capiz who were recruited by a relative to work in a plantation. They found out that their salaries had been advanced to the recruiter. In Sabah, they stayed in the bunkhouse which was located fifteen kilometres from the plantation – they walked this distance every day. They had no days off, no contact with their families, and most of the time they ate dried fish. They also had no way of asking for assistance. Therefore, when they heard they might be sold to another plantation, they ran away. They walked for five days and nights until they reached Lahad Datu, from where they took a boat to Tawi-Tawi where they stayed until they were able to secure assistance from the Capiz provincial government. Some of them later discovered that they had caught malaria.

Thus, the working and living conditions of unauthorized migrants in Sabah are not conducive to promoting good health. Their vulnerabilities to health hazards are exacerbated by the lack of support from the state, as well as support structures that the Philippine government usually extends to its nationals overseas. There is no Philippine consulate or embassy in Sabah. This means that there are no labour attaches, welfare officers or Filipino resource centres to address the needs and concerns of Filipino nationals, particularly migrant workers. Migrant NGOs are either not visible or are absent in Sabah.[34] The Catholic Church, a source of support and assistance to migrants in many settings, is at an early stage of its mission with migrants (Lim 2002). When asked who they approached for assistance, migrants said that they helped each other.

In general, unless questions about health were asked, unauthorized migrants did not readily volunteer any information on health matters. As mentioned above, migrants' narratives of their life in Sabah were framed mostly in terms of their unauthorized status. Yet the findings from the 2000b study on unauthorized migration reveal that next to work-related problems, health problems were the second most common problem among them. Nevertheless, most of their health problems were not major in the sense that they had to seek professional consultation. When they became sick, migrants sought out their relatives, friends and employers/supervisors for assistance (Battistella and Asis 2003b: 105).

No interviewees in the 2000a study reported having experienced major health problems. In the 2000b study, a few mentioned malaria, which they attributed to drinking dirty water. The only interviewee who had sought hospital care was a female refugee, an identity card holder, who said that her husband's hospitalization was covered by the Sabah government. None of the other interviewees had sought government-run facilities to address their health problems. Instead, the data suggest that Filipinos in Sabah turned to alternative health care providers or practices.

There are various forms of medical assistance sought by unauthorized migrants. They may seek out private medical practitioners who do not require them to present their identify cards. When Ning fell ill, for example, she went to a private doctor. She paid RM40 for the consultation and was absent from work for five days which meant RM50 was deducted from her salary. Those who work in the plantations can access basic health care provided by the employer. One plantation near Sandakan had a clinic staffed by a nurse.[35] Some of those who worked in construction projects mentioned that their employers provided assistance when they became ill. However, some former construction workers also shared that their employers did not extend any assistance when they became sick. Traditional healers are another possible source of health care. One interviewee in Bongao was a healer who had spent some time in Sabah. He said that he had treated many patients in the *kampong* who had sought his help for various illnesses (Interview, 2000a study).

An official at the Provincial Health Office in Tawi-Tawi mentioned that some of their patients were former migrants to Sabah. While it would be difficult to ascertain whether they became sick in Sabah, he did acknowledge that their poor working and living conditions in Sabah could render them susceptible to diseases or injuries. This was corroborated by the account of an interviewee in the 1997 study who had to return to the Philippines when he became ill. He had worked as a construction driver in Sabah and earned good money, until he found out that he had kidney problems. He returned to the Philippines without receiving any medical assistance from his employer; he was an unauthorized migrant.

Concerns about 'illegals' burdening the state's health care system seem to be more imagined than real. The data from the three studies generally point to migrants assuming responsibility for their health care. Data from the Health Department, on the other hand, show that foreigners (although it was not clear whether they were legal or unauthorized) accounted for a sizeable share of outpatients and inpatients in government hospitals in Sabah (Kurus 2002). The public's concerns that foreigners are putting a strain on the health system have led to increasingly restrictive policies.

While their status as unauthorized migrants in Sabah serves as some kind of a template in defining their life chances, including their health conditions, the data also suggest that vulnerability to health risks is not the same for all unauthorized migrants. Rather, the situation of Filipinos in Sabah points to several variables as factors that could modify the relationship between unauthorized migration and health outcomes.

As noted above, one of the sources of the diversity of Filipino communities in Sabah derives from their region of origin. Ethnicity is an important variable, since it is related to migrants' stock of resources and constraints, particularly in accessing services, including health care. 'Traditional' migrants (Tausugs, Samals, Badjaus and Yakans), for example, have the advantage of established social networks given their long history of cross-border mobility. Social networks could provide vital information and assistance to migrants. Other Filipino migrants may lack social capital, but they may have more human capital, particularly better incomes, which could promote their well-being.

This is something to consider when exploring the risk-taking behaviour of migrants. Contrary to the perception that migrants tend to engage in risky behaviours (deriving from anonymity and freedom from social control in the new environment), there is also the possibility that migrants may be more cautious due to the lack of familiarity, lack of resources or social control by the community. In the case of Sabah, traditional migrants tend to move with other family members, which provides migrants with social support which, in turn, reduces the likelihood of migrants undertaking risky practices. Even where family members are absent, social networks can exercise some pressure on migrants to behave according to socially accepted norms.

Migration to Sabah may be arranged independently (this is true for 'traditional' migrants) or it may be arranged with the mediation of agents (including traffickers). The former does not involve costs to migrants, and migrants are also basically free of demands and conditions imposed by agents and brokers. Either alone, or with the help of their social networks, they can get jobs or search for better wages or working conditions. In times of illness, they can access health care depending on the resources available to them. At the other extreme are those who have been trafficked; their choices are severely constrained (for example, the experiences of the twenty-three migrants who were trafficked to work in the plantation).

The importance of occupation was suggested by the 1997 study on HIV vulnerability in the Mindanao-Sabah migration system which pointed out that the occupation of migrant workers could predispose them to varying degrees of vulnerability to HIV infection. Workers in the sex-related sectors rank very high in this regard. Aside from their potential exposure to infection, they do not receive support (especially from their employers) to engage in health-promoting practices. In the case of other workers, employers may be one source of assistance in times of illness – we found this to be quite common among construction workers, although this is highly dependent on the generosity of employers.

The experiences of unauthorized Filipino migrants in Sabah highlight powerful structural factors that render migrants vulnerable to many risks, including health problems, which severely constrain their capacity to address their problems. Although not all unauthorized migrants are equally vulnerable to health problems, their lack of legal status and the lack of rights associated with it seem to be *the* major factor in defining their working and living conditions. Because they are unauthorized, Filipinos do not exist as far as the state (both Sabah and the Philippines) is concerned. Because they are unwanted, solutions such as the *Ops Nyah* (literally, 'get rid') operations have become a standard approach in removing the 'problem'. Because they are unprotected, unauthorized migrants such as the Filipinos in Sabah are basically left to themselves in dealing with problems with employers, or health, or authorities. The situation is further complicated by the lack of alternative services available to migrants. Disqualifying migrants from accessing critical social services (including access to education for migrant children) further marginalizes them. In Sabah and other receiving areas or countries, there is an increasing resolve to deal decisively with unauthorized migration. Without due attention to the factors that render migrants unauthorized, unwanted and unprotected, durable solutions may not yet be in sight.

Notes

1 This term is used in this chapter to refer to what is commonly labelled as 'illegal migration'. Unauthorized migration or irregular migration refers to migration that departs from norms governing the exit, entry, stay and work of non-nationals. Unlike the term 'illegal migration', unauthorized migration or irregular migration recognizes that there are many actors involved in the process and it does not unduly stigmatize migrants (Battistella and Asis 2003a: 11).

2 Wong and Teuku Afrizal (2003: 168–9) provide an interesting discussion about the evolution of the terminology referring to unauthorized migration in Malaysia, from illegal immigrant (*pendatang haram*) to illegals, and from the mid-1990s, 'aliens'. The Indonesian-designated term is *migran gelap* (migrants in the dark or shadow migrants).

3 Since the mid-1990s, the estimate of the migrant worker population in Malaysia has hovered around 1.2 million, of whom at least 60 per cent are unauthorized. In 2002, estimates of 600,000 to 1 million irregular migrants were reported in the media; legal migrant workers were said to number 750,000 (*Asian Migration News* 31 May 2002, 30 June 2002). Indonesians are the largest group of both legal and unauthorized migrant workers in Malaysia. Filipinos figure as another major group of unauthorized migrants in Malaysia – they are mostly in Sabah.

4 The Philippines' claim to Sabah dates back to 1704 when North Borneo was ceded to the Sultan of Sulu by the Sultan of Brunei in return for the former's help in quelling a rebellion. In 1878 the Sultan of Sulu signed a lease for Sabah with the British Company. Subsequent events complicated the situation – the transformation of Sabah into a British crown colony, the granting of Philippine independence (which included the Sultanate of Sulu) and the creation of the Federation of Malaya. The Philippines filed a claim for Sabah in 1962 and actively pursued it in the 1960s and 1970s – it has been relatively dormant since then (Battistella *et al.* 1997). The 2002 repatriations prompted a call to pursue the claim, but this was not followed up and the controversy died down.

5 Mindanao is one of three major island groupings in the Philippines (the others are Luzon and the Visayas). Although the region is identified as home to Filipino-Muslims, non-Muslims (including *Lumads* or indigenous populations) are the majority of Mindanao's total population. Commonly referred to as southern Philippines, Mindanao is subdivided into six administrative regions: Western Mindanao (Zamboanga Peninsular), Northern Mindanao, Southern Mindanao (Davao Region), CARAGA, Soccsksargen and the Autonomous Region of Muslim Mindanao (ARMM). Unless otherwise stated, in this chapter I refer to Zamboanga City, Basilan, Sulu and Tawi-Tawi as part of Western Mindanao based on a geographical criterion. Note that administratively, Basilan, Sulu and Tawi-Tawi are part of ARMM. Zamboanga City is not identified as a stronghold of any of the ethnolinguistic groups in Western Mindanao. Because of its location, it receives migrants from other parts of Western Mindanao each time there are peace and order problems in the surrounding areas.

6 Recruitment agencies require applicants to take a medical examination in designated clinics, the results of which cannot be used by the applicant to present to another agency.

7 In the Philippines, HIV/AIDS-antibody testing is voluntary, as provided in the Philippine AIDS Prevention and Control Act of 1998. But if it is required by the country of destination, the Philippine government cannot force the country of destination to uphold the Philippine law. Eventually, it is the applicant who decides whether or not to submit to testing.

8 The medical examination for seafarers is guided by AO no. 176, s. 2000 (Standard Guidelines for Conducting Medical Fitness Exam for Filipino Seafarers). In

the case of seafarers, the recruitment agencies shoulder the cost of the medical examination. (Please note that in the Philippines, the term 'recruitment agencies' is used to refer to agencies recruiting 'land-based' workers, while manning agencies is the term used to refer to those recruiting and hiring seafarers.)

9 The actual cost of the medical examination is much less. The huge margin between how much workers are charged and what is actually paid to the clinic is another source of revenue for placement agencies.

10 To ensure that the medical examination is done according to standards, the Department of Health, through the Bureau of Licensing and Registration, is responsible for accrediting and monitoring medical clinics and hospitals. As of the time of writing, there are some 140 accredited clinics and hospitals nationwide, of which 85 per cent are in Metro Manila. Despite the monitoring system, there are reports of anomalies by clinics and hospitals, such as requiring tests or procedures that are unnecessary.

11 Provisions of the standard employment contract which pertain to health include the following responsibilities of the employer to the worker: providing free emergency medical and dental services and facilities, including medicines; personal life and accident insurance in accordance with host government and/or Philippine government laws without cost to the workers; and repatriation if the worker is unable to continue working due to work-related or work-aggravated injury or illness. In areas adjudged by the Philippine government as war risk areas, a war risk insurance of not less than P100,000 is provided by the employer at no extra cost to the worker.

12 Legal overseas Filipino workers enrol in the Overseas Workers Welfare Administration – Medicare (fee of P900) which entitles themselves and their dependants to hospital and medical benefits. The Medical Care for Overseas Filipino Workers (OFWs) was created under Executive Order 195 on 13 August 1994. A move to transfer the medicare programme from OWWA to the Philippine Health Insurance Corporation and other changes was under discussion in September 2003.

13 Since the seminar has to cover many topics, health matters may not receive much attention. For seafarers, however, the seminar includes a module on HIV/AIDS.

14 I wish to thank the anonymous reviewer who raised this question which led me to consider how unauthorized migrants overcome the disadvantages of their legal status in managing their health problems.

15 BIMP–EAGA is the largest subregional co-operation in Southeast Asia. Its formation capitalized on the traditional ties among the neighbouring areas. It includes the Sultanate of Brunei; Central Kalimantan, North Sulawesi, South Sulawesi, East and West Kalimantan, South Kalimantan, Central Sulawesi, Southeast Sulawesi, and Irian Jaya in Indonesia; Sabah, Sarawak and Labuan in Malaysia; and Mindanao and Palawan in the Philippines. Except for Brunei, all the other member-areas are among the least industrialized areas in their respective countries. The latest estimate puts the population of BIMP–EAGA at 57.5 million.

16 The north-to-south movement of people in Mainland Southeast Asia was also accepted until recent evidence indicated 'a more stable, indigenous development during the region's prehistory than had previously been assumed' (Wright 1990: 44).

17 The study on unauthorized migration in Southeast Asia found cross-border flows to be the most common form of unauthorized migration in the region. These are largely undocumented flows (that is, not involving the use of travel and/or work documents) between geographically and culturally similar areas – for example the accessibility of Malaysia is a factor in the concentration of Indonesians there;

also, the 1,600 km shared by Burma and Thailand allows for the cross-border flows. The Philippines has a rather unique pattern of unauthorized migration because it involves more distant and diverse destinations. As such, unauthorized migration from the Philippines often includes the use of travel and/or work documents, which are necessary in gaining admission to another country. An exception to this pattern is the migration from Mindanao to Sabah, which hews closer to undocumented cross-border movements (see Battistella and Asis 2003a). As Figure 1 shows, the Philippines is relatively isolated from the rest of Southeast Asia, except for Western Mindanao, which is close to Sabah.

18 These are not the sole factors that lead to unauthorized migration. In analysing unauthorized migration in Southeast Asia, we argued that there are determinants and facilitating factors operating in the area of origin and area of destination that lead to international migration. Both unauthorized and legal migrations have similar determinants, except that unauthorized migrants have access to unauthorized channels, while legal migrants have access to legal channels (see Battistella and Asis 2003a).

19 Palawan is another exit/entry point to and from Sabah. Interestingly, this route has not captured the popular imagination in the same way that the southern back door has.

20 These ethnolinguistic groups are dominant in Western Mindanao: Tausugs ('people of the current') are the majority in Sulu; the Yakans, in Basilan; and the Samals and Badjaus (both are the original inhabitants of Tawi-Tawi) form smaller groups throughout the area. Other Muslim groups are found in other parts of Mindanao.

21 Travel time between Taganak Island and Bongao, the capital of Tawi-Tawi, is ten hours by motor launch; travel time between Taganak Island and Sandakan is less than an hour.

22 This is a motorized medium-sized hydrovessel made out of local wood.

23 Some respondents implicated the participation of border personnel in this practice (see, for example, Bunny's account on pp. 129 and 131).

24 A review of the BIMP–EAGA's performance and prospects may be found in ADB (1996), Chavez-Malaluan (1998) and Yussof and Mohd (2003), among others.

25 Residents in Tawi-Tawi generally thought that taking the ferry was more expensive and bothersome compared to the familiar and accessible *kumpit* system. The overnight trip from Bongao to Sandakan is faster than the ferry service. Going legal would mean getting a passport and travelling to Zamboanga City, the departure point.

26 According to Kurus (2002:70): 'Based on the explanation provided by the Prime Minister's Department, the Filipino refugees were allowed entry into Malaysia and allowed to stay and work without any time limitation in Sabah and Labuan under the Passport Order Act (exemption) (2) (amendment) 1972, which was signed by the Minister of Home Affairs on September 12, 1972'.

27 PBS founding President Datuk Joseph Pairin Kitingan said that when the Party started in 1984, among others, it strove to be the 'voice of the people who wanted issues such as illegal migrants here to be resolved'. The Party has consistently called for the elimination of 'phantom voters'. In 1990, it split from the Barisan Nasional but recently re-entered the fold (http://wwww.pbs-sabah.org/pbs3/html/party/background.html, accessed 9 September 2003).

28 It is interesting to note that Filipino settlers also expressed similar sentiments. Many of them also said that the people in West Malaysia do not fully understand Sabah.

29 Well-off Filipino interviewees in Sabah expressed frustration at being lumped together with the 'illegals' who 'drag down' the rest of the law-abiding Filipino population.

30 When Sabah was a British protectorate, the British authorities had taken to importing foreign workers to support its economy. When Sabah joined the Federation of Malaysia in 1963, it continued to attract migrants from the neighbouring areas (Kurus 1999: 282). The arrival of refugees from Mindanao in the 1970s was notable for the numbers involved and the possibility that they would settle in Sabah. Subsequent arrivals from Mindanao (and later from other parts of the Philippines) were migrants in search of work.

31 In order to have better estimates, Professor Azizah Kassim of the Universiti Sabah Malaysia said that there is a need to define irregular migrants. She asked whether those who have lived in Sabah for half a decade would still be considered 'illegal' (*Asian Migration News* 15 July 2003).

32 The names are fictitious.

33 The proliferation of fake identity cards (IC) is a persistent problem in Sabah and in Malaysia in general. One of the things that employers ask of job applicants is the IC.

34 In Thailand, for example, in addition to government clinics, there are NGOs which provide health care to migrants (see Amarapibal *et al.* 2003).

35 The clinic, however, did not provide family planning services. Although the workers were provided with bunkhouses, water supply was intermittent, which could pose health problems.

References

Amarapibal, A., Beesey, A. and Germershausen, A. (2003) 'Irregular migration into Thailand', in G. Battistella and M.M.B. Asis (eds) *Unauthorized Migration in Southeast Asia*, Quezon City: Scalabrini Migration Center.

Asian Development Bank (ADB) (1996) *East Asian Growth Area: An Investigative Report*. Manila: Asian Development Bank.

Asian Migration News (various issues) (2002–2004). Online, available at: www.smc.org.ph/amnews/amnarch.htm.

Battistella, G. and Asis, M.M.B. (2003a) 'Southeast Asia and the specter of unauthorized migration', in G. Battistella and M.M.B. Asis (eds) *Unauthorized Migration in Southeast Asia*. Quezon City: Scalabrini Migration Center.

Battistella, G. and Asis, M.M.B. (2003b) 'Irregular migration: the underside of the global migrations of Filipinos' in G. Battistella and M.M.B. Asis (eds) *Unauthorized Migration in Southeast Asia*. Quezon City: Scalabrini Migration Center.

Battistella, G., Asis, M.M.B. and Abubakar, C. (1997) *Migration from the Philippines to Malaysia: An Exploratory Study*. Report submitted to the International Organization for Migration. Quezon City: Scalabrini Migration Center.

Chavez-Malaluan, J.J. (1998) 'From "Flying Geese" to "Cog and Wheel": some issues on subregional economic zones', *ARENA*, 14, 1: 9–32.

Daily Express (2003) 'Call for drastic action to deal with illegals', 24 April. Online, available at: www.dailyexpress.com.my/news.cfm?NewsID=18438 (accessed 8 August 2003).

Department of Health (1999) 'Revised rules and regulations governing accreditation of medical clinics and hospitals and the conduct of medical examination of overseas workers and seafarers', Administrative Order 85-A, Series 1990. Manila: Department of Health.

Gaur, S. and Saxena, P.C. (2004) 'Indian migrant workers in Lebanon and their access to health care', *Asian and Pacific Migration Journal*, 13, 1: 61–88.

Human Development Network (HDN) and the United Nations Development Programme (UNDP) (2002) *Philippine Human Development Report 2002*. Makati City: HDN and UNDP.

Isarabhakdi, P. (2004) 'Meeting at the crossroads: Myanmar migrants and their use of Thai health care services', *Asian and Pacific Migration Journal*, 13, 1: 107–26.

Kurus, B. (1999) 'Migrant labor in Sabah', *Asian and Pacific Migration Journal*, 7, 2–3: 281–95.

Kurus, B. (2002) 'Undocumented immigrants in Sabah: reality, trends and response', *Asian Migrant*, 15, 3: 68–73.

Lim, D. (2002) 'Pastoral care for migrants and refugees: challenges and obstacles for the local church in Sabah', *Asian Migrant*, 15, 3: 74–6.

Mindanao Economic Development Council (MEDCo) (2004). 'Cheers to BIMP–EAGA at 10'. Online, available at: www.medco.gov.ph/medcoweb/bimpeagaanniversary.asp (accessed 11 September 2004).

OneOcean (n.d.). 'Philippine Turtle Islands'. Online, available at: oneocean.org/ambassadors/track_a_turtle/tihpa/pti.html (accessed 8 August 2003).

United Nations Development Programme (UNDP) (2000) *Assessing Population Movement and HIV Vulnerability: Brunei–Indonesia–Malaysia–Philippines Linkages in the East ASEAN Growth Area*. Bangkok: Scalabrini Migration Center and UNDP.

Warren, J.F. (2003) 'The Balangigi Samal: the global economy, maritime trading and diasporic identities in the nineteenth-century Philippines', *Asian Ethnicity*, 4, 1: 7–30.

Wong, D. and Teuku Afrizal T.A. (2003) '*Migrant Gelap:* irregular migrants in Malaysia's shadow economy', in G. Battistella and M.M.B Asis (eds) *Unauthorized Migration in Southeast Asia*. Quezon City: Scalabrini Migration Center.

Wright, L. (1990) 'Iranon migration in Mindanao and Borneo', in R. Reed (ed.) *Patterns of Migration in Southeast Asia*. Berkeley: University of California Press.

Yussof, I. and Mohd, Y.K. (2003) 'Human resource development and regional cooperation within BIMP–EAGA: issues and future directions', *Asia-Pacific Development Journal*, 10, 2: 41–56.

8 Migration, differential access to health services and civil society's responses in Japan

Keiko Yamanaka

Introduction

In November 1991, Jeevan Shrestha, a 22-year-old migrant worker from Nepal, fell sick in Japan.[1] Four months later in a clinic in the city of Hamamatsu, Shizuoka Prefecture, he was diagnosed as suffering from a fatal deficiency in the arterial valves of his heart. To save his life, Dr Hiroshi Taniguchi, who examined the patient, decided that he needed to undergo open-heart surgery to replace his failing arterial valves with artificial ones. The surgeon estimated the cost of surgery at more than five million yen (US$45,000), plus more than two million yen (US$18,000) for hospitalisation, medication and treatment.[2] Shrestha was an unauthorised (illegal) visa-overstayer who was ineligible for public health insurance, as a result of which public funds would not be available to him. This posed a serious dilemma for Dr Taniguchi and his colleagues in the general hospital where Shrestha was admitted. By then, local media had featured his story, attracting a great deal of public attention in the city and its vicinity. In response, sympathetic citizens sent donations, a total of 600,000 yen (US$5,400), to contribute to his surgery, while a group of volunteers coordinated efforts to assist him. Doctors and citizens alike wished to save Jeevan's life but faced a dilemma: who was going to pay his huge medical costs?

This example epitomises growing gaps between Japan's policy of neglect towards the social welfare of immigrant workers, on the one hand, and the many incidents of sickness and injuries that befall them, on the other. By law, Japan prohibits unskilled foreigners from gaining employment. The unprecedented economic boom of the late 1980s, however, demanded more labour than was legally available, to which some employers responded by ignoring the law. As a result, a large influx of guest workers was drawn to the country. By the mid-1990s, Japan was home to more than half a million guest workers and their families from Asia and Latin America. The rapid growth of the immigrant population was a result of a number of 'back doors' that had opened in response to the contradictory policy through which unskilled migrant workers were able to enter as legal tourists, students, entertainers, company trainees, long-term residents, and relatives of

Japanese nationals (Yamanaka 1993; Cornelius 1994; Tsuda and Cornelius 2004).

As the Shrestha case highlights, from the beginning of this influx, migrants' access to health services has become a focus of intense policy concerns and contentious public debate (Miyajima and Higuchi 1996). Most of these newcomers, both authorised and unauthorised, are excluded from enrolling in public health insurance plans. However, a high incidence of work-related injuries and personal illness among immigrant workers, especially those who are unauthorised, has drawn public attention to the inequality and injustice experienced by this most vulnerable of categories of workers in Japan.

Japanese citizens' responses to mounting immigrant health problems have been spontaneous and in sharp contrast to the continuing neglect by government and by industries that depend heavily on immigrant labour. Since the early 1990s, a few committed citizens – often members of such groups as labour unions, professional association and religious organisations – have assisted immigrants in meeting their health and medical needs. They have also advocated for the rights of immigrants as residents, workers and human beings, in public debates, publications and campaigns, and in lobbying elected officials and government offices. In many industrial cities where immigrants have settled, frequently with their families, citizen activists have recognised that their lack of access to health care is an urgent and alarming community problem to which activists have responded with determination (Hamamatsu NPO Network Centre 2001). This clearly suggests that recent global immigration has energised Japan's civil society, and motivated concerned citizens to fill the gap between outdated governmental policies and the social problems that they cause for immigrant workers.

In this chapter, I discuss how grass-roots forces have responded to governmental neglect in health care for increasing numbers of uninsured foreign residents in Hamamatsu. By way of introduction, I first briefly describe Japan's experience during the post-Second World War era with immigrant workers, and how the government handled the human rights of these workers and their descendants. The description begins with regional comparisons between Europe and Asia as they reveal different ways in which the rights of immigrant workers are conceived and administered in the two regions.

Citizenship rights of migrants in Europe and Asia

The influx of immigrants, refugees and asylum seekers into Western Europe since the 1950s has resulted in significant expansion of membership rights for foreigners there (Soysal 1994). The ways in which these foreigners have been incorporated into the host countries' legal and institutional structures vary depending on their national and political histories. However, over the years many European nation-states have granted substantial social, economic and political rights to authorised immigrants. As a result, by the 1990s citizenship ceased to be a determining factor in their eligibility for most public services, economic activities and political participation at local levels.[3] In 'post-national' Europe, therefore, immigrants' rights are increasingly determined by residence, not by citizenship.[4] Throughout the region, among the public services available to foreign residents, access to inexpensive health and medical care is fundamental to the levels of public health and social welfare in the nation. Inclusive legal stipulation notwithstanding, however, institutional and cultural discrimination embedded in the bureaucracy, labour markets and medical organisations often prevents ethnic minority populations from full access to their statutory benefits. To alleviate this problem, immigrant associations and citizens' organisations provide immigrants with formal and informal services and advocate their rights as local citizens (see e.g. Piper 1998).

Since the end of the Second World War, the United Nations and other international and regional organisations have endorsed treaties and conventions that guarantee basic rights and welfare to refugees and immigrants and their families in their host countries (Soysal 1994; Sassen 1998). These treaties and conventions commonly define human rights as the universal and inalienable rights of individual persons, regardless of personal attributes such as sex, nationality, ethnicity and class. In recent decades, under the leadership of the European Union and as a consequence of globalised migration, Europe has been increasingly integrated – deterritorialised – as a result of which individual rights have become increasingly important. Therefore, in 'the postnational model, universal personhood replaces nationhood; and universal human rights replace national rights. . . . Hence, the individual transcends the citizen' (Soysal 1994: 142). For many civil activists and organisations struggling to enhance immigrant rights, these international laws have provided much needed moral principles and legal tools with which to combat legal, institutional and cultural barriers in the face of governmental neglect and public indifference (Gurowitz 1999, 2001).

Outside Europe and a few other Western nation-states, however, the concept of universal personhood has attracted little attention. In Asia, Africa and Latin America, many governments are building the nation-state and constructing nationhood. In many East and Southeast Asian countries, legacies of colonialism, civil war and poverty have long divided the population along ethnic and regional lines (see e.g. Maidment *et al.* 1998). As a result, many governments continue to direct most of their efforts to solidifying the foundation of the nation-state (Castles and Davidson 2000). Although migrant workers provide a pool of inexpensive and flexible labour, indispensable to economic development, these governments tend to maintain rigid control over their borders and to draw sharp distinctions between citizens and non-citizens (see e.g. Wong 1997; Yamanaka 1999). Some Asian governments emphasise 'Asian values' to uphold collective values, dismissing individual rights as 'Western values' (Bauer and Bell 1999). Some are yet to ratify international laws that require signatories to raise labour standards and treat non-citizens as equal to citizens.

In Asia, immigration laws typically permit unskilled migrants to work for a short period of time in the host country. The laws usually stipulate rules regarding their admission and exit, but give little attention to immigrant rights during their residence (Battistella 2002). The laws also prohibit family members from joining their migrant relatives at their destination. Such official neglect of human rights for migrant workers has created serious problems in their daily lives, since they are regularly exposed to labour accident and injury, abuse and violence, illness and depression, and epidemics such as AIDS/HIV. When illness or injury befalls them, migrants often receive neither medical treatment nor payment of wages earned. They have little recourse but to accept deportation.

Despite frequent violations of such human rights at the hands of bureaucrats and employers, citizens of most Asian labour importing countries remain unaware of these problems, and therefore pay little attention to the plight of foreign labourers. The mass media tend to reinforce negative public images of them by repeatedly disseminating inaccurate information (see e.g. Chin 2003). As a result, throughout Asia, non-government organisations and civil groups are the only agents dedicated to the protection of migrants' rights (Yamanaka and Piper 2003). In response to the critical importance of personal and public well-being, Asian NGOs have commonly worked to increase public awareness of health issues, while providing health services and health training for migrant men and women (see e.g. Van Beelen *et al.* 2001, Verghis and Fernandez 2001).

Japan and exclusive immigration policy

Japan occupies a unique position in its immigration policy and citizenship rights for migrant workers, intermediate between post-national Europe and post-colonial Asia. On the one hand, it stands out as having been the only Asian coloniser in the pre-Second World War period and 'the only long-standing Asian democracy' in the post-Second World War period (Castles and Davidson 2000: 196). Having achieved rapid economic reconstruction and development, post-war Japan resembles Western Europe in its high level of technological sophistication and high living standards, its large middle-class population, and its solid foundation as a nation-state. On the other hand, unlike its European counterparts, Japan has adopted an official policy of admitting only skilled foreigners, while unofficially admitting more than half a million unskilled, de facto guest workers (as discussed below). Because they lack citizenship, most of these newcomers are denied rights to family reunification and access to public services. In its exclusive policy on immigration and citizenship, Japan resembles its Asian post-colonial neighbours. The contradiction between Japan's advanced capitalism and its obsolete immigration policies requires ethno-historical explanation dating back to its nation-state building era of more than one hundred years ago (Yamanaka 2004a, b).

In addition to these newcomers, Japan, like all nation-states, includes within its population a number of long-term and indigenous ethnic minorities with distinct histories and cultures. This is a result of nation-building efforts since the 1860s that forced ethnic minorities – Koreans, Chinese, Ainu, Okinawans, *Burakumin* and others – to assimilate into the national culture and polity dominated by the majority Japanese population (Lie 2001). Hence, a brief account of foreigners' rights to public services, especially their eligibility for national health insurance and an old-age pension, provides a clue to official ideologies and discourses of exclusion based on nationality throughout the post-Second World War era. It also reveals the unprecedented impact that international laws, ratified by the Japanese government, have had in leading the government to amend discrimination, laws and bureaucratic procedures. Furthermore, it shows the crucial role played by such non-government actors as labour unions, professional and religious organisations, and immigrants' associations in urging the government to eradicate discrimination based on nationality.

Citizenship rights of oldcomers

In May 1947, Japan's New Constitution took effect. The American Occupation Forces drafted its original text in English, using the words 'all of the people' as the subject of the sentences concerning rights and privileges. From its inception, these words generated confusion and concern regarding the rights of non-national residents in Japan, most of whom were former colonial citizens from Korea (estimated at approximately half a million). This is because the Japanese text of the Constitution has translated 'all of the people' as *subete kokumin* (all nationals), thereby neglecting the presence of resident non-Japanese, while manifesting a strong exclusive nation-state ideology (Takafuji 1991: 8). According to Article 14 of the Constitution, foreign nationals are guaranteed equality: 'All of the people [in Japanese, 'all nationals'] are equal under the law and there shall be no discrimination in political, economic or social relations because of race, creed, sex, social status or family origin.' However, throughout the post-war era, there has been no legislation in Japan that specifically bans discrimination based on race, nationality or ethnicity.

The Constitution's Chapter III, 'Rights and Duties of the People', defines a wide range of citizenship rights for Japanese nationals. Specifically, Article 25 on the right of survival (*seizonken*) states: 'All people [in Japanese, 'all nationals'] shall have the right to maintain the minimum standards of wholesome and cultured living. In all spheres of life, the State shall use its endeavours for the promotion and extension of social welfare and security, and of public health.' Following this stipulation, in 1958 the National Health Insurance Law was established to launch the National

Health Insurance (*Kokumin Kenko Hoken*) to cover those nationals who were not already covered by employee health insurance; that is, self-employed, farmers, retirees and their dependants. Similarly, in 1959 the National Pension Law was instituted to mandate that all adult nationals participate in the National Pension Programme (*Kokumin Nenkin*). For those who are employed, welfare pension insurance (*Kosei Nenkin*) has been in place since 1941 to provide health insurance and an old-age pension programme.

In these earlier periods of Japan's efforts to provide social welfare and security to all nationals, a serious debate arose surrounding the question of whether or not these public programmes should be extended to all non-Japanese. Most legal scholars of the time considered that the right of survival could not be denied to foreigners, although, as noted above, in Article 25 'all people' refers to Japanese nationals (Takafuji 1991: 8). In 1950, the Supreme Court recognised this right even for illegal entrants to the country. However, three years later, under the deepening tensions of the Cold War, the same Court ruled that the constitutional guarantee for right of survival could be applied to foreigners '*except* those rights that were interpreted to be guaranteed only for Japanese nationals because of their nature' (Takafuji 1991: 8; emphasis added). According to this ruling, rights of access to public services in health, social welfare and social security were considered to be preserved only for Japanese. Consequently, after this decision, foreign nationals were categorically denied these rights despite the fact that foreigners, most of whom were former colonial citizens, were taxpayers and law-abiding residents.

The differential exclusion of foreigners from national health insurance and social security insurance continued until 1985 when Japan ratified the United Nations Convention and Protocol Relating to the Status of Refugees (Refugee Convention hereafter). By the 1970s, Japan was an economically developed country under increasing pressure from the world community to meet international standards of human rights. In 1981, the government agreed to sign the Refugee Convention in response to international pressure to join other nations in admitting Indochinese refugees in far larger numbers than Japan had reluctantly proposed to admit. The Convention required its signatories to guarantee equal treatment of foreign nationals in access to social welfare and social security services. Obliged by the Convention, the government finally revised domestic laws to include all registered foreign nationals in its National Pension Programme (1982), and to include all foreign nationals residing in the country for more than one year in the National Health Insurance Programme (1986).

In addition to the Refugee Convention, since the late 1970s the Japanese government has ratified a series of international human rights laws. These include the International Covenant on Civil and Political Rights (1979), the International Covenant on Economic, Social and Cultural Rights (1979), the Refugee Convention (1981), the Convention on the Elimination of All Forms of Discrimination Against Women (1985), the Convention on the

Rights of Children (1994), and the Convention on the Elimination of All Forms of Racial Discrimination (1995). According to Gurowitz (1999), the universal mandates of these conventions provided Japanese and Korean activists with much needed moral support in the face of widespread public indifference to their causes. They also greatly helped the activists bolster their arguments against the government which, up until then, had been extremely reluctant to extend citizenship rights to foreign nationals.

The 1982 and 1986 provisions, allowing former colonial citizens ('old-comers'; 700,000 Koreans and 200,000 Chinese) access to the National Health Insurance Programme and National Pension Programme, epitomised activists' successful application of international norms in their campaign for the human rights of immigrants (Gurowitz 1999). Moreover, after decades of intense domestic and international pressure, in 1993 the Japanese government finally eliminated the notorious practice of fingerprinting permanent residents as a requirement for official registration. Such progress suggests that international norms can have an impact on domestic policy in the context of changing social consciousness for human rights (Gurowitz 1999). In this instance, by integrating international norms with domestic causes, civil activists played a crucial role in raising public awareness and pressuring policy-makers to grant the basic rights of those individuals who had been denied rights as a result of legal, institutional and cultural barriers (see e.g. Tegtmeyer Pak 2000: 263).

Citizenship rights of newcomers

By 1990, the arrival of no less than 200,000 unskilled migrant workers redefined citizenship rights for 'newcomers' in Japan. The majority came from neighbouring Asian countries to take those jobs shunned by Japanese. Most were able to enter the country by overstaying short-term visas issued to tourists and some other categories of visitors, allowing them to remain for no more than ninety days (Morita and Sassen 1994). In response to the over-stayers in December 1989 the Ministry of Justice revised the Immigration Control and Refugee Recognition Law, while retaining its principle of limiting foreign labour to skilled occupations (Yamanaka 1993, 1996; Cornelius 1994; Weiner and Hanami 1998). The revised law, which took effect in June 1990, instituted a criminal penalty for employers found to have hired illegal foreign workers. In an effort to maintain the supply of unskilled workers while stemming the flow of unskilled 'foreigners', the same law created a long-term, unrestricted residence visa exclusively for foreign descendants of Japanese emigrants (*Nikkeijin*).

This resulted in an influx over the next five years of more than 200,000 authorised *Nikkeijin* workers and their dependants from Latin America; Brazil in particular (Yamanaka 1996, 2000a; Roth 2002; Tsuda 2003). During the same period, there were an estimated 300,000 unauthorised visa overstayers. Consequently, by 1995 Japan was hosting more than half a million unskilled migrant workers in industrial and metropolitan cities throughout the country (Cornelius 1994; Tsuda and Cornelius 2004). Because their visas were short-term or expired, the majority of the newcomers were ineligible for most public social services, including national health insurance. Although all workers, regardless of legal status, are required by law to participate in the labour accident insurance and the unemployment insurance programmes, foreign workers rarely did so. This was despite the fact that the majority worked in small factories and on construction sites requiring long hours with few or poor safety measures. In times of recession, they were the first to be laid off.

Among the long-term resident Japanese Brazilians and their families, those who had arrived in the early 1990s were able to enrol in the national health insurance. This, however, came to an abrupt halt in 1993 when the Ministry of Health and Welfare sent an internal memorandum to all local governments, ordering them not to accept applications by foreign workers – mostly *Nikkeijin* – for national health insurance (Roth 2002: 72). Instead, the Ministry urged the governments to advise foreign workers to enrol in social security insurance (*shakai hoken*) through their employers because they were defined as employed workers. By law, employers and employees are required to enrol and contribute funds equally to Social Security Insurance (a combination of health insurance and old-age pension). This administrative advice resulted in a sudden decline in the proportion of *Nikkeijin* who had health insurance coverage because most *Nikkeijin* workers were young and healthy and did not intend to stay in Japan for longer than a few years. They did not want to pay the expensive insurance instalments that would not be reimbursed fully upon their return to their home country.[5] Most employers of these workers were small-scale labour brokers who were also financially motivated to ignore such costly programmes. According to anthropologist Joshua Roth, who conducted research in Hamamatsu, by the mid-1990s only 20 to 30 per cent of the city's 7,000 *Nikkeijin* had enrolled in the National Health Insurance Programme (Roth 2002: 72).

Civil activism for newcomers' access to inexpensive health care: a case study

On 31 March 1992, Nepalese patient Jeevan Shrestha underwent successful heart surgery in Hamamatsu, and by late April was able to leave hospital. Under this extraordinary circumstance, the Immigration Bureau granted him a special visa to stay one more year in Japan for follow-up treatment. While receiving treatment, he lived in an apartment with the help of Japanese volunteers and his family members who were also working in Japan. A year later on 31 March 1993, he left for Nepal in good health. Upon his departure, it was estimated that the total cost of his surgery, hospitalisation and medication had amounted to 6,500,000 yen (US$58,600). Of that cost, 1,700,000 yen (US$15,300) was paid by donations from the Nepalese migrant community and local Japanese sympathisers (*Chunichi Shimbun* 1993).[6] The balance of 4,800,000 yen (US$43,200) remained unpaid until the city government finally settled it from its undesignated budget (Taniguchi 2000). This set an unfortunate precedent for thousands of uninsured foreigners in the city, since most hospitals began to avoid them, even when they were in a critical condition (see e.g. *Shizuoka Shimbun* 1992). At the same time, the Shrestha incident had an unexpected and long-lasting impact on voluntary activism among concerned citizens of Hamamatsu and its neighbouring cities.

Hamamatsu is a city of half a million in western Shizuoka Prefecture, 257 kilometres southwest of Tokyo which, together with its neighbours, hosts the headquarters for several major automobile, motor cycle and musical instrument companies such as Suzuki, Honda, Yamaha and Kawai Piano. Since the late 1980s, acute labour shortages concentrated among small subcontractors of these large companies drew large numbers of *Nikkeijin* workers and their families, mostly from Brazil (Yamanaka 2000a; Ikegami 2001a; Roth 2002). Hamamatsu alone, for example, received 3,448 Brazilians in 1990, tripling to 11,182 ten years later. The industrial area also became home to many unauthorised Asian workers (Yamanaka 2000b).

This unprecedented influx of immigrant workers caused dedicated Japanese citizens of the area to organise a variety of community services for foreign workers and their families (Ikegami 2001a). The citizens' activities were intended to meet the needs of foreigners whose lack of citizenship and cultural familiarity denied them access to information on labour laws, inexpensive health care and housing, legal rights and political representation. As early as 1990, a citizens' voluntary group, the *Herusu no Kai* (Hamamatsu Overseas Labourers Solidarity), had begun to provide consultation and mediation services for foreign workers confronting labour disputes, job injuries, unpaid wages and problems of social welfare (Ikegami 2001b: 260–1). Other citizens who were dedicated to education formed networks and groups to teach immigrant children the Japanese language in community centres (Takeuchi 2001).

In 1992 when Shrestha was hospitalised, the regional media prominently reported his case. As a result, his story reached many local citizens, who then raised questions about health care for foreign workers, especially those who were unauthorised. It also sent a strong message to medical professionals and municipal administrators about the financial and social risks of the failure to establish policies regarding uninsured patients. Yet neither party took action or came forward with solutions. Hamamatsu citizens' responses to the mounting problems were spontaneous and determined, and in sharp contrast to the continuing refusal of the local government to grant foreign workers access to inexpensive health care. By the mid-1990s two voluntary associations – *Grupo Justiça e Paz* and the Medical Association for Foreigners – emerged in order to meet the health needs of foreign workers and their dependants.

Underlying the emergence of these community organisations in Hamamatsu is a growing emphasis on self-governance at the grass roots throughout Japan. Arising from the ashes of the 1995 Kobe Earthquake, this new civil society movement stresses voluntarism, public interest, non-profit-making and non-government organisation (Tajiri 2001: 19). In contrast to traditional community activism serving the interests of a specific neighbourhood, the new community activism addresses broad societal concerns including education, health, ageing, disability, environment, immigration, human rights and so on. Under increasing budgetary constraints, coupled with the rapidly ageing population, local governments are inclined to delegate policy projects to non-government organisations (Sakuma 2001: 147–8). In an age of decentralisation of state power, the partnership between local governments and non-government organisations has been consistent with the interests of the national government.

Grupo Justiça e Paz

Grupo Justiça e Paz evolved from the congregation of a Catholic church in Hamamatsu, many members of which were Brazilian and Peruvian nationals of Japanese descent. Before coming to Japan, the church's priest, Father Edvaldo Yano, had long worked for the relief and emancipation of the poor in Brazil. Upon arrival in Hamamatsu, with help of *Nikkei* and Japanese members, he provided a variety of services for the city's *Nikkei* populations, including advice, consultation, documentation, and negotiation with bureaucrats and employers. In April 1995, the Catholic group formed *Grupo Esperança* to provide a weekly meal service for the city's homeless, mostly Japanese, who lived in parks and public spaces near Hamamatsu Central Station. In 1993 when Jeevan Shrestha fell ill, Masako Yoshino, a member of the church and *Herusu no Kai*, coordinated communication between the citizen's group and the Nepalese migrant community. As the numbers of the city's uninsured foreigners increased, Father Yano and his group received increasing numbers of calls seeking assistance with health and medical problems (Ameishi *et al.* 1998). In response, in March 1997, Yano, Yoshino and other church members formed *Grupo Justiça e Paz* (Group of Justice and Peace, hereafter *Grupo*). They began a public campaign urging the Hamamatsu municipal government and the City Assembly to expand membership of the national health insurance to include the city's foreign residents.

In their petition of 12 March 1997 addressed to the Mayor, *Grupo* documented the serious problems faced by uninsured foreign populations (Ikegami 2001b: 234–5). The petition cited examples of medical emergency relief funds that had been established for foreigners by other local governments and demanded five items: (1) that the Hamamatsu city government conduct a study on enrolment of foreign residents in the National Health Insurance Programme and accept their applications; (2) that the Social

Security Agency ease procedures for foreigners to pay social security premiums in order to promote their participation in the Social Security Insurance Programme; (3) that the city government create a new health insurance plan designed specifically for foreign worker populations; (4) that the Shizuoka prefectural government launch a medical emergency relief fund for uninsured foreigners; and (5) that the city government establish a new department responsible for providing foreigners with public services. A month later, *Grupo* submitted a similar petition to the Shizuoka prefectural governor (Ikegami 2001b: 235; *Shizuoka Shimbun* 1997).

In May, the Hamamatsu government and the City Assembly's Social Welfare and Health Committee replied, stating that the city's policy was to follow the national government's guidance, which was not to admit foreign workers into the National Health Insurance Programme (Ikegami 2001b: 237–8). On the remaining four items, policy-makers made it clear that they had no intention of implementing any of *Grupo*'s proposals. Instead, they replied that they would forward the petition to the relevant agencies of the Prefecture and the national government. The hostile attitude of the City Assembly Committee was reflected in an incident which occurred during its discussion of *Grupo's* proposals. In frustration, provoked by continued questioning on the part of *Grupo* members, one Assembly member made anti-foreigner comments.[7] This incident was widely publicised, igniting angry protests from the Brazilian community and Japanese activists (Ikegami 2001b: 238–242; Roth 2002; 73).

In June, the Shizuoka prefectural government and prefectural assembly committees reached conclusions similar to those of its Hamamatsu counterparts (Ikegami 2001b: 235–6). They replied to *Grupo* that it would be unnecessary to institute new policies to promote health access for foreign nationals, or to issue new administrative guidelines to local municipal administrations because, according to prefectural policy-makers, local administrators were already following national ministerial guidance.

Dissatisfied with the indifferent replies by the local governments, *Grupo* continued its campaigning, collecting signatures from both foreigners and citizens in promotion of foreigners' access to inexpensive health care (Ikegami 2001b: 242). In June and July, with 5,500 signatures, *Grupo* resubmitted its petition to the Hamamatsu Mayor, who gave answers similar to those given by the City Assembly's Social Welfare and Health Committee in May (Ikegami 2001b: 243–6). Shortly after this *Grupo* ended its campaign in defeat. Over the next two years, the main members of *Grupo* formed a new group, *Justiça e Fraternidade*, and continued their activities, primarily collecting data on uninsured foreign residents in forty cities throughout Japan with a high concentration of foreigners. In late September 1999, Yano, Yoshino and two others visited the Ministry of Health and Welfare in Tokyo, requesting eleven items regarding participation of foreigners in public health insurance plans and relief measures for uninsured patients. In response, the Ministry's officials repeated its policy of not admitting foreign workers into the National Health Insurance Programme (*Shizuoka Shimbun* 1999).

Grupo's short-lived but intense campaign on behalf of foreigners' rights thus ended with little progress in changing public policies at either the local or national levels. However, *Grupo* set a precedent for civil action in Hamamatsu where few citizens had been aware of the lack of inexpensive health care for foreigners. Its actions demonstrated that non-citizens do demand rights and equality as residents, and many citizens support their cause. *Grupo*'s campaign sent a clear message to the nation, from a city with a high concentration of foreign workers, that legalised inequality and discrimination based on nationality would not be tolerated in silence.

Medical aid for foreigners in Hamamatsu (MAF)

Ever since Jeevan Shrestha had arrived at his clinic in 1992, Dr Taniguchi, the heart surgeon who had operated on him, had been increasingly concerned about foreigners' health care issues (Sakai and Ikegami 2001: 260). As a medical professional and supporter of *Herusu no Kai*, he had already been made aware of the problems met by uninsured foreigners, especially those who were unauthorised. The moral and financial dilemma that the case brought to Dr Taniguchi's attention when he decided to operate on Shrestha, led him to seek an alternative way to meet the medical needs of foreigners. In 1996, when the Rotary Club Central celebrated its tenth anniversary, Taniguchi, a member, suggested that the Club sponsor an event in which free medical checkups would be provided to uninsured foreigners by volunteer citizens, interpreters and professional groups of doctors, nurses, nutritionists and technicians. The Rotary Club adopted his proposal and decided that it would sponsor the event in cooperation with *Herusu no Kai* (Ikegami 2001b: 260–2).

After two months of intense preparation, this unprecedented charity programme took place on a Sunday in October 1996 in a large public hall near the Hamamatsu Central Bus and Railway Station. Despite the absence of medical facilities at the location, and no experience of such services, 209 volunteers of diverse capacities (twenty-three doctors and dentists, eighteen nurses, sixteen technicians, fifty-four interpreters and ninety-eight others) and nationalities (Japanese, Brazilian, Peruvian, Filipino and others) successfully carried out the medical screening of 259 foreigners, adults and children, authorised and unauthorised, from ten countries (Brazil, Peru, The Philippines, Nepal and others) (Hamamatsu Rotary Club Central 1997). The examinations available were in the areas of internal medicine, paediatrics, mental health and dentistry. Six tests were given to all examinees: blood test, blood type, blood pressure, urine, electrocardiogram and chest X-ray. This was the beginning of annual free checkup programmes which continue to take place.

The success of the first such event greatly encouraged its leaders and volunteers, and attracted considerable media and public attention as well. In June 1997, the same people formed Medical Aid for Foreigners in Hamamatsu (MAF), to continue their service to Hamamatsu's increasing numbers of uninsured foreign residents, with the former leader of *Herusu no Kai* as its Chairperson (Ikegami 2001b: 263). Its main goals included the provision of annual checkups, the elimination of administrative barriers to foreigners' access to inexpensive health care, and the promotion of employers' participation in the Social Security Insurance Programme in order to increase enrolment among their foreign employees. MAF's initial success won the trust and interest of the city's major general hospitals. In the same year, the second free checkup programme was held, this time in the Hamamatsu Red Cross Hospital where all necessary facilities were available. Two years later the venue changed to the larger Enshu General Hospital, where the programme has been held ever since.

The October 2002 checkup programme served 587 foreigners of diverse nationalities (MAF Hamamatsu 2003). A total of 327 Japanese and non-Japanese volunteered their services: fifty-two doctors, forty-six nurses, forty-four technicians, eighty interpreters and 105 others (MAF Hamamatsu 2003). In 2003, gynaecology, otolaryngology and plastic surgery were added to the four medical specialities that had previously been provided. The test items increased from six to eight by adding checks for breast and uterine cancers. The number of interpreters also increased to facilitate communication with speakers of Portuguese, Spanish, English, Tagalog, Indonesian and Bengali. Over the years, the list of financial and technical support organisations and individuals has increased from eight to forty-five, and includes the city's major general hospitals, associations of medical workers and specialists, rotary and lions clubs, an international exchange organisation, a printing company, a voluntary association of Brazilian medical workers, and a variety of individual donors.

Despite this support and the success MAF has experienced in serving mostly uninsured foreigners, it faces significant ongoing problems, not least of which is financial. Donations comprise its major source of income (*Shizuoka Shimbun* 2001). As a voluntary association, it does not receive public subsidies. Consequently, its annual income has fallen far short of its annual expenses. MAF's 1998 balance sheet, for example, reports an income of 1,140,000 yen (US\$9,500) derived entirely from donations, and expenses totalling 973,400 yen (US\$8,112) (MAF Hamamatsu 1999).[8] The expenses do not include the actual cost of checkups: 8,000 yen (US\$67) per examinee (Taniguchi 2000). Most medical expenses are covered by the free services of volunteers and supporting establishments. That is, medical doctors, technicians, nurses and nutritionists provide their expertise and labour; medical supply companies contribute supplies; medical laboratories perform tests; and hospitals bear overhead expenses – all without charge (*Shizuoka Shimbun* 2001). As the programme attracts more applicants each year, MAF's leaders wonder and worry about how long this kind of generous charity can continue.

Another major problem faced by MAF is medical. After each checkup, MAF reports the results to each examinee. Because the programme is only intended to check a person's physical condition, those patients whose results indicate health problems are expected to visit a physician or hospital at their own expense to receive diagnosis and treatment. Over the years, however, MAF has learned that the majority do not do so (Hamamatsu NPO Network Centre 2001: 6). This is primarily because it is hard for temporary foreign workers to take a day off from work (hospitals are closed on weekends) or, in the absence of insurance, they are unable to pay for treatment. This is a problem beyond MAF's capacity to address.

Despite these many problems and constraints, MAF leaders and volunteers remain optimistic and energetic, organising regular meetings throughout the year to administer and plan activities. MAF is determined to continue its service until such time as all foreigners in the city (and the country) are covered by public health insurance. In recent years, MAF has received requests for assistance from civil groups in other cities which have become interested in launching similar programmes. Thus MAF's message is spreading, and its precedent is being followed beyond the vicinity of Hamamatsu.

In an interview, when asked why MAF provides the checkup service free of charge, Dr Taniguchi answered:

> In Japan, there are many kinds of free medical services provided to Japanese: for infants, school children, pregnant women, company employees, adults and the elderly. However, there is nothing like that for foreigners. This is why we do this free. If our MAF services were to be provided on a fee basis, our responsibility to the examinees would be greater than we can bear as volunteer providers. There is much to do. We need to raise funds for the programme while preparing it throughout the year. As a way to do [this], we try to appeal to citizens' social conscience because to do so is to fulfil the foreigners' human rights. I believe this works better in the long run than making it political by blaming the administration for its failure to provide medical services for them.
>
> (Taniguchi, Interview, 26 January 2000)

Forces from below and above

Grupo and MAF have adopted very different methods and strategies to approach the same goal. The former, composed of ordinary citizen and immigrant activists, has briefly, but directly, challenged the local and national governments for the expansion of foreigners' rights. In contrast, MAF has for years been successful in mobilising large numbers of medical professionals, mainstream institutions and prominent citizens to participate in its once-a-year voluntary activities. It has also been successful in attracting increasing numbers of uninsured foreigners to take advantage of its free annual checkups.[9] In January 2001, a local business and management research institute awarded MAF the International Exchange Service Award for its distinguished activities for the welfare of the local population (*Herusu no Kai* 2002).

These two civil groups' actions for foreigners' rights demonstrate the kind of civilian mobilisation that has recently attracted academic interest as a form of 'governance from below'. This concept refers to the actions of ordinary citizens when they participate in 'the exercise of power in a variety of institutional contexts, the object of which is to direct, control, and regulate activities in the interests of people as citizen voters, and workers' (Robinson 1996: 347; see also Falk 1993). Where this kind of democratic governance within a nation-state is enacted across ethnic boundaries as a result of migration, which is, in turn, a result of the globalised economy, scholars find that it comprises a form of political 'transnationalism from below'. Therein, coalitions of men and women of various nationalities, ethnicities and classes exercise power for common goals transcending national boundaries (Guarnizo and Smith 1998; Portes 1999; Lister 1997).

Despite such an impressive solidarity of forces from below as is exhibited in Hamamatsu by *Grupo* and MAF, it remains to be seen what impact that solidarity will actually have on policy-makers in the city and national governments. As of summer 2003, no major change had been announced to improve foreigners' access to inexpensive health care. The traumas of Japan's colonial past account in large part for the post-war government's indifference to citizenship rights for foreigners in Japan. In contrast, Japan's recent history of civil protest indicates that great strides have been taken by immigrant organisations and their Japanese sympathisers in expanding equality and justice for the oldcomer populations. Embracing international human rights laws as their moral guide, civil activists of diverse nationalities in Japan have tenaciously challenged legal, administrative and institutional barriers embedded in the bureaucracy, labour markets and social customs.

The legacy of activists' struggles continues in the many cities where thousands of newcomers, including their families, have settled. In Hamamatsu, the struggles to eliminate barriers to equality and justice for foreigners and other minorities have taken a variety of forms. One such form is vividly illustrated by a specific instance of legal action that had the effect of integrating an international law into the legal defence of a local Hamamatsu woman victimised by 'racial' discrimination. In 1998, the city attracted unwanted national mass media attention which focused on a discrimination lawsuit brought by Brazilian journalist Ana Bortz, who sued a local jewellery storeowner, T. Suzuki, who, she charged, had attempted to expel her from his store on the basis of her nationality (Maeda 1998). In the absence of any applicable law, Bortz cited the authority of the International Convention on Eliminating All Forms of Racial Discrimination, which had been ratified by Japan since 1995. A year later, the District Court judge who presided over the case astounded the city and the nation when he ruled that the plaintiff had suffered discrimination because of her race and nationality, and ordered the defendant to pay her full compensation. The judge had decided, in agreement with Bortz's argument, that in view of the absence of any domestic anti-discrimination law, and in view of Japan's ratification of the International Convention, the Convention's provisions must serve as the standard by which racial discrimination was to be determined, prohibited and adjudicated (French 1999).

A study of the social impact of this unprecedented Court ruling on Hamamatsu citizens' perceptions of foreigners suggests that many thought that defendant Suzuki did the 'right' thing in order to protect his business from a foreigner, whom he regarded as a potential criminal on the basis of her nationality (Yamanaka 2003a). They also expressed vague fears generated by the fact that a foreigner had dared to challenge Japanese authority and hegemony in a Japanese city. The long-term impact of the Court ruling on public perceptions and attitudes towards ethnic minorities is yet to be seen in Hamamatsu. It is evident, however, that the issues of racial equality and human rights which were emphasised in the Court ruling remain a remote concept in the minds of the majority of citizens who have never doubted their entitlement to ethnic (national) dominance in Japan, thus accepting discrimination against foreigners as reasonable (Yamanaka 2004b).

Conclusion

This case study of Hamamatsu indicates that concerned citizen civil activists' responses to the emergence of newcomer communities have been spontaneous, determined and positive. In sharp contrast, the responses of local and national governments have been ambiguous at best, while the general public has ranged from indifferent to hostile. This suggests that Japan's path to building a multi-ethnic community is clearly destined to be rocky, with much social and political tension embedded in the process of challenging the status quo in which the dominant group and its government have long maintained power, privilege and hegemony. In this process, international law and grass-roots activism have proven to be two important new forces in expanding the ideological landscape and legal framework of citizenship in Japan. More research is necessary before it will be possible to predict how, and to what extent, these two forces – international law applied from 'above' and grass-roots activism emerging from 'below' – can bring significant change to the conservative 'middle' – the government and the public.

Notes

1 All personal names in this chapter are pseudonyms. Information about the Jeevan Shrestha case was obtained from newspaper articles (*Chunichi Shimbun* 1992a, 1992b, 1993; *Yomiuri Shimbun* 1992a, 1992b), interviews with knowledgeable civil activists in the city, and an interview with Shrestha in Kathmandu, Nepal on 24 December 1996.
2 In 1993, the exchange rate was 111 yen per US dollar.
3 Soysal (1994) analyses the incorporation of guest workers and their dependants (who have become permanent residents) into the states' political structures in six Western European countries since the 1970s (Sweden, the Netherlands, Germany, France, Switzerland and Britain). Guest worker populations from non-EU member countries, such as Turkey and those in North Africa, continue to face many forms of exclusion from the political, social and economic institutions in receiving countries (Piper 1998).
4 It should be stressed that this is limited mostly to citizens of EU member countries. In fact, in 1990 the Constitutional Court of Germany ruled out the extension of local voting rights to non-EU nationals who had not become naturalised citizens. A Turkish citizen, for example, would be prohibited from voting, whereas a French citizen would have this right (Yamanaka and Piper 2003: 9).
5 Since April 1996, a partial refund of the first three years of instalments that foreigners invested in the National Social Security Plan has been made available to them when they leave Japan (Roth 2002: 73–4). This, however, benefits few Brazilians as they rarely enrol in the Plan. If they were to do so, they would receive no return beyond their first three years because many work for more than three years, while continuing to be required to pay into the Plan.
6 For information about Nepalese labour migration to Japan and their community activities, see Yamanaka 2000b and 2003b.
7 The Assembly member stated: 'If they [foreigners] say they do not want to work here unless they can participate in the health insurance program, let them go home as they wish' (Ikegami 2001b: 239).
8 The exchange rate was 120 yen per dollar in 1998.
9 Since 2002, MAF has launched a programme in which pupils in the city's Brazilian schools are able to receive a free checkup. In July 2003, 672 pupils from the three Brazilian schools and one Peruvian school received a free checkup (*Herusu no Kai* 2003: 1–2).

References

Ameishi, T., Oda, T., Tsubraya, A., Fujita, M. and Yamauchi, S. (1998) *Hamamatsu-shi Kyoju Gaikokujin heno Iryo to Hoken Teikyo ni tuite*. Study Report by Nursing Students. Hamamatsu: author.

Battistella, G. (2002) 'International Migration in Asia vis-à-vis Europe: An Introduction', *Asian and Pacific Migration Journal*, 11, 4: 405–14.

Bauer, J.R. and Bell, D.A. (eds) (1999) *The East Asian Challenge for Human Rights*. Cambridge: Cambridge University Press.

Castles, S. and Davidson, A. (2000) *Citizenship and Migration: Globalization and the Politics of Belonging*. New York: Routledge.

Chin, C.B.N. (2003) 'Visible Bodies, Invisible Work: State Practices Toward Migrant Women Domestic Workers in Malaysia', *Asian and Pacific Migration Journal*, 12, 1–2: 49–73.

Chunichi Shimbun (1992a) 'Hoken Kikazu: Shinzo no Nanshujutsu Seiko', 3 April.

—— (1992b) '*Minasama no Zeni de Inochi Tasukarimashita: Nepal Seinen ga Reijo*', 7 April.

—— (1993) '*Motto Itaikedo . . . Jutsugo 1 nen, Munen no Kikoku*', 1 April.

Cornelius, W.A. (1994) 'Japan: The Illusion of Immigration Control', in W.A. Cornelius, P.L. Martin and J.F. Hollifield (eds) *Controlling Immigration: A Global Perspective*, Stanford, CA: Stanford University Press.

Falk, R. (1993) 'The Making of Global Citizenship', in J. Brecher, J.B. Child and J. Cutler (eds) *Global Visions*. Boston, MA: South End Press.

French, H.W. (1999) '"Japanese Only" Policy Takes Body Blow in Court', *New York Times*, International Edition, pp. A1, A14, 15 November.

Guarnizo, L.E. and Smith, M.P. (1998) *Transnationalism from Below*. New Brunswick, NJ: Transaction Publishers.

Gurowitz, A. (1999) 'Mobilising International Norms: Domestic Actors, Immigrants, and the Japanese State', *World Politics*, 51, 3: 413–45.

—— (2001) 'Migrant Rights and Activism in Malaysia: Opportunities and Constraints', *Journal of Asian Studies*, 59, 4: 863–88.

Hamamatsu NPO Network Centre (2001) *Daremoga Kenkouni Kuraseru Shakai wo Mezashite*. Hamamatsu: author.

Hamamatsu Rotary Club Central (1997) *Gaikokujin no tameno Muryo Kenko Sodan to Kenshinkai 1996, Hokokusho*. Hamamatsu: author.

Herusu no Kai (2002) '*Hamamatsu Gaikokujin Iryo Enjokai ga "Kokusai Koryu Korosho" wo Jusho*', *Herusu no Kai News Letter* 125 (March).

—— (2003) '*Burajirujin, Perujin Gakko Kenshin no Kekka ni tsuite*', *Herusu no Kai News Letter*, 136 (October).

Ikegami, S. (ed.) (2001a) *Brajirujin to Kokusaika suru Chiiki Shakai: Kyoju, Kyoiku, Iryo*. Tokyo: Akashi Shoten.

—— (2001b) '*Gaikokuseki Teijusha to Iryo Hosho:* Grupo Justiça e Paz *no Chinjo ga Toikaketa Mono*', in S. Ikegami (ed.) *Brajirujin to Kokusaika suru Chiiki Shakai: Kyoju, Kyoiku, Iryo*. Tokyo: Akashi Shoten.

Lie, J. (2001) *Multiethnic Japan*. Cambridge, MA: Harvard University Press.

Lister, R. (1997) *Citizenship: Feminist Perspectives*. New York: New York University Press.

Maeda, T. (1998) 'Brazilian Files Discrimination Suit: Apology, 1.5 Million Yen Sought', *Japan Times*, 3 September, P.3.

Medical Aid for Foreigners (MAF) Hamamatsu (1999) *Heisei 11 nendo Hamamatsu Gaikokujin Iryo Enjokai Sokai*. Hamamatsu: author.

—— (2003) *Heisei 14 nendo Gaikokujin Muryo Kenshinkai Hokokusho: 2002*. Hamamatsu: author.

Maidment, R., Goldblatt, D. and Mitchell, J. (eds) (1998) *Governance in the Asia-Pacific*. London: Routledge.

Miyajima, T. and Higuchi, N. (1996) '*Iryo, Shakai Hosho: Seizonken no Kanten kara*', in T. Miyajima and T. Kajita (eds) *Gaikokujin Rodosha kara Shimin he: Chiiki Shakai no Shiten to Kadai kara*. Tokyo: Yuhikaku.

Morita, K. and Sassen, S. (1994) 'The New Illegal Immigration in Japan 1980–1992', *International Migration Review*, 28, 1: 153–63.

Piper, N. (1998) *Racism, Nationalism and Citizenship: Ethnic Minorities in Britain and Germany*. Aldershot: Ashgate.

Portes, A. (1999) 'Conclusion, Towards a New World: The Origins and Effects of Transnational Activities', *Ethnic and Racial Studies*, 22, 2: 463–77.

Robinson, M. (1996) 'Governance', in A. Kuper and J. Kuper (eds) *The Social Science Encyclopaedia*. London: Routlege.

Roth, J.H. (2002) *Brokered Homeland: Japanese Brazilian Migrants in Japan*. Ithaca, NY: Cornell University Press.

Sakai, M. and Ikegami, S. (2001) '*Hamamatsushi ni okeru Gaikokujinn Iryo no Torikumi: Burajirujin wo Chusin ni*', in S. Ikegami (ed.) *Brajirujin to Kokusaika suru Chiiki Shakai: Kyoju, Kyoiku, Iryo*. Tokyo: Akashi Shoten.

Sakuma, T. (2001) '*Global Jidai no Borantia Katsudo wo Kangaeru*', in the Editorial Committee *Borantia Hakusho 2001*. Tokyo: Nihon Seinen Hoshi Kyokai.

Sassen, S. (1998) *Globalization and Its Discontents*. New York: The New Press.

Shizuoka Shimbun (1992) '*Gaikokujin no Iryohi Fuharai Kyuzo: Kuni ya Ken ni Yobo*', 16 October.

—— (1997) '*Chiji, Kenkai Gichou ni Chinjo: Gaikokuseki Teijusha no Iryo Hosho*', 17 April.

—— (1999) '*Kokuho Kanyu wo Kibo: Hamamatsu no Nikkeijinra Koseisho ni*', 1 October.

—— (2001) '*Hamamatsu Hoshiki Motto Shitte: Gaikokujin Kenshin*', 11 November.

Soysal, N.S. (1994) *Limits of Citizenship: Migrants and Postnational Membership in Europe*. Chicago, IL: University of Chicago Press.

Tajiri, K. (2001) '*Kininaru Borantia to NPO no Kankei*', in the Editorial Committee *Borantia Hakusho 2001*. Tokyo: Nihon Seinen Hoshi Kyokai.

Takafuji, A. (1991) '*Gaikokujin Rodosha to Wagakuni no Shakai Hoshosei*', in Shakai Hosho Kenkyujo (ed.) *Gaikokujin Rodosha to Shakai Hosho*. Tokyo: University of Tokyo Press.

Takeuchi, H. (2001) '*Kominkan Katsudo ni Miru Nihonjin Shaki to Burajirujin Shaki no Setten*', in S. Ikegami (ed.) *Brajirujin to Kokusaika suru Chiiki Shakai: Kyoju, Kyoiku, Iryo*. Tokyo: Akashi Shoten.

Taniguchi, H. (2000) Interview, Hamamatsu, 26 January.

Tegtmeyer Pak, K. (2000) 'Foreigners Are Local Citizens, Too: Local Governments Respond to International Migration in Japan', in M. Douglass and G.S. Roberts (eds) *Japan and Global Migration: Foreign Workers and the Advent of a Multicultural Society*. London: Routledge.

Tsuda, T. (2003) *Strangers in the Ethnic Homeland: Japanese Brazilian Return Migration in Transnational Perspective*. New York: Columbia University Press.

Tsuda, T. and Cornelius, W.A. (2004) 'Japan: Government Policy and Immigrant Reality', in W.A. Cornelius, T. Tsuda, P.L. Martin and J.F. Hollifield (eds) *Controlling Immigration: A Global Perspective* (2nd edn). Stanford, CA: Stanford University Press.

Van Beelen, N. (documentor), and I. Fernandez and S. Verghis (eds) (2001) *Report, Mobility and HIV/AIDS: Strengthening Regional Interventions*, Satellite Symposium on Mobility and AIDS: 5th International Congress on AIDS and the Pacific. Kuala Lumpur: author.

Verghis, S. and Fernandez, I. (eds) (2001) *Regional Summit on Pre-departure, Post-arrival and Reintegration Programs for Migrant Workers*. Kuala Lumpur: CARAM Asia.

Weiner, M. and Hanami, T. (eds) (1998) *Temporary Workers or Future Citizens? Japanese and U.S. Migration Policies*. New York: New York University Press.

Wong, D. (1997) 'Transience and Settlement: Singapore's Foreign Labour Policy', *Asian and Pacific Migration Journal* 6, 1: 135–67.

Yamanaka, K. (1993) 'New Immigration Policy and Unskilled Foreign Workers in Japan', *Pacific Affairs* 66, 1: 72–90.

—— (1996) 'Return Migration of Japanese-Brazilians to Japan: The *Nikkeijin* as Ethnic Minority and Political Construct', *Diaspora*, 5, 1: 65–97.

—— (1999) 'Illegal Immigration in Asia: Regional Patterns and a Case Study of Nepalese Workers in Japan', in D.W. Haines and K.E. Rosenblum (eds) *Illegal Immigration in America: A Reference Handbook*, Westport, CT: Greenwood Press.

—— (2000a) '"I Will Go Home, but When?": Labour Migration and Circular Diaspora Formation by Japanese Brazilians in Japan', in M. Douglass and G.S. Roberts, (eds) *Japan and Global Migration: Foreign Workers and the Advent of a Multicultural Society*. London: Routledge.

—— (2000b) 'Nepalese Labour Migration to Japan: From Global Warriors to Global Workers', *Ethnic and Racial Studies*, 23, 1: 62–93.

—— (2003a) 'A Breakthrough for Ethnic Minority Rights in Japan: Ana Bortz's Courageous Challenge', in M. Morokvasic-Muller, U. Erel and K. Shinozaki (eds) *Gender and Migration: Crossing Borders and Shifting Boundaries*, Volume 1, International Women's University Series. Opladen, Germany: Verlag Leske+Budrich.

—— (2003b) 'Transnational Activities for Local Survival: A Community of Nepalese Visa-overstayers in Japan', *Kroeber Anthropological Society Papers*, 89/90: 147–67.

—— (2004a) 'Citizenship and Differential Exclusion of Immigrants in Japan', in B.S.I. Yeoh and K. Willis (eds), *State/Nation/Transnation*. London: Routledge.

—— (2004b) 'Citizenship, Immigration and Ethnic Hegemony in Japan', in E.P. Kaufmann (ed.) *Majority Groups and Dominant Minorities: Conceptualising Dominant Ethnicity*. London: Routledge.

Yamanaka, K. and Piper, N. (2003) 'An Introductory Overview', *Asian and Pacific Migration Journal*, 12, 1–2: 1–19.

Yomiuri Shimbun (1992a) '*Chiryohi 1,000 man Chuni: Fuho Shuro, Nanbyo no Nepal Seinen*', 3 April.

—— (1992b) '*Nepal Seinen ni Shien Zokuzoku: Fuho Shurochu ni Shinzo Shujutsu*', 7 April.

9 An institutional approach towards migration and health in China[1]

Xiang Biao

Introduction

There are at least eighty-five million rural–urban migrants in mainland China. These migrants face great health risks, yet are not covered by any medical care scheme. In linking migration and health, particularly in studies of migration and HIV, most of the existing literature adopts a 'behavioural approach', which assumes that certain migrant behaviour patterns are disease-prone. Moving away from this perspective, I propose an 'institutional approach' which holds that migrants face particular health problems because of their position in the established social system. In terms of policy recommendations, the behavioural approach tends to focus on policies that directly affect migrants' health behaviours, while the institutional approach draws attention to the fact that medical policies are embedded in various other institutions, and accordingly stresses the importance of examining the relationship between different institutional arrangements and identifying realistic strategies for policy change.

This chapter comprises three sections. The first describes the basic health problems migrants face and demonstrates that their major problems stem from institutional factors. The second section explores why migrants are not included in China's new medical care system that supposedly aims to cover as many people as possible. A political economy analysis of China's medical care system and migrants' relation to this system suggests that the common policy option put forward by researchers and advocates – to include migrants in formal medical care – may not be feasible in the near future. Accordingly, the final section proposes alternative policy recommendations based on the above analysis.

My analysis is based on documentary research conducted from December 2002 to May 2003, and fieldwork investigation in Beijing and Zhejiang, China, in April and May 2003. During fieldwork I interviewed approximately twenty researchers and policy-makers in the Ministry of Health, State Council Development Research Centre, Ministry of Labour and Social Security, Beijing University, Qinghua University and other institutes. I visited Dashila Street in Xuanwu District, Beijing, and its community health centres, and interviewed officials, health workers and residents there. Apart from this, I also draw on information from my earlier long-term research with migrants in China starting in 1992, particularly my work on migrants in the Pearl River Delta and on a migrant community in southern Beijing.

Migrant health problems

My documentary research and fieldwork suggests that work-related injuries and illness constitute the biggest health problem for migrants. Lack of access to existing health services is the immediate cause of many other health problems. Although HIV and other sexually transmitted diseases (STDs) among migrants have attracted considerable attention, I contend that the linkage between migration and STDs is yet to be established. There is insufficient evidence that migrant behaviour is particularly risky. Migrants may be vulnerable to HIV, just as they are vulnerable to other diseases, but this is due to their position in society rather than their particular behaviour.

Work-related injuries and illness

It is widely known that migrants typically fill job positions with high health risks, which are sometimes referred to as '3-D' (Dirty, Dangerous, and Dead-end/Difficult/Demanding) jobs. According to a survey by Tan and her associates in 1994, approximately one-third of the migrant workers in six cities in the Pearl River Delta believed that their health had been affected by their working conditions, particularly by noise, dust and poison (Tan 2002a: 145). An incomplete record reported 12,000 work-related accidents and more than eighty deaths a year in Shenzhen city, Guangdong province, most of the victims being migrants, and a report on work-related accidents and illness submitted to the State Council by the then State Economy and Trade Commission in July 2000 identified migrant workers as the main victims of all work-related health problems (cited in Tan 2002a: 145).

Work-related health problems include work-related injury and work-related illness. Tan estimates that approximately 1 to 2 per cent of all male migrant workers had work-related injuries (Tan 2002b: 255). In 1998, Shenzhen reported 189 cases of migrant workers becoming physically disabled at work, mostly losing fingers or arms (Yu 2001). Far more serious than injuries are chronic diseases acquired from work. Besides the high treatment costs, victims of chronic diseases often develop symptoms only after leaving the workplace, making it difficult at times to determine which factory should be held responsible for the disease. Pneumoconiosis, which has been found among many construction workers, is one such disease.[2] In one case, a group of more than 200 migrant workers in Liaoning province worked eight hours daily without any protective measures in air containing

97.6 per cent silicon dioxide. Silicon dioxide cannot be dissolved in the body and, in the long term, clogs the alveoli. By October 2001, ten workers had died and 192 had lost their ability to work due to severe pneumoconiosis. They had no medical insurance and a few had to stop medical treatment half-way through due to financial difficulties (see Dai *et al.* 2001; Pan Guanglin *et al.* 2001). By the end of 2000, China had 425,000 patients suffering from this disease (Gao 2001).

Benzene poisoning is probably the most common serious chronic disease found among female migrant workers. Large numbers of benzene poisoning cases have been reported in Beijing, Hebei, Zhejiang, Fujian and Guangdong, particularly in garment, shoe or suitcase factories which often use cheap glue with high compositions of benzene. As many as 70,000 female workers, mostly migrants, at shoemaking factories in Putian county, Fujian province, had suffered from benzene poisoning by 1996 (see Chen 1996). Benzene damages the blood and nervous system and can be fatal; it is also very difficult to cure.

Female workers are also particularly vulnerable to fire accidents, because they are often housed in congested dormitories without basic safety equipment. In 1991, seventy-two female migrant workers lost their lives in a fire in Dongguan city, Guangdong province. A fire in Shenzhen in 1993 killed eighty-seven migrants, eighty-five of whom were women (Tan 2002c: 312). Based on a collection of news reports, Tan (2002d: 79) estimated that by 1994, at least 300 female migrant workers in the Pearl River Delta alone had been killed in fires. Given the highly hazardous working conditions of migrant workers, it is not an exaggeration to suggest that China will have a huge group of unproductive, and even ill former migrants in one or two decades if their health problems are not addressed in a timely manner.

Lack of access to existing health services

Many migrant health problems are caused by lack of access to existing health services. According to a survey of migrant women in the Jiading District of Shanghai, nearly 80 per cent of the migrant women hoped to have regular reproductive health checks, but only 17.3 per cent did so, and 55.5 per cent did not know where they could obtain help regarding family planning (Zhang 1999: 56). Surveys conducted during the period from 1996 to 1998 on migrant children in Wuhan, Hubei province, reported that only 30.2 per cent had been properly vaccinated as required by the government, and 4.0 to 7.5 per cent of migrant children had not been vaccinated (Xiao Xianwu 1999: 761).

The key reason for the lack of this capacity is financial difficulty. Very few migrants have access to financial assistance for medical treatment. In a survey conducted in Chengdu city, Sichuan province, and Shenyang city, Liaoning province, no migrants had medical insurance (Guan and Jiang 2002: 258). According to a survey in Beijing, 93 per cent of migrants who had fallen ill had not received any payment for their medical expenses from their employers (Wen *et al.* 2003, cited in Huang and Pieke 2003).

Due to financial constraints, access to major hospitals is a key problem for most migrant workers in China. Guan Xingping and Jiang Miaoyi (2002) reported that migrants could afford only about 100 RMB for medical treatment a month, but one consultation in a big hospital for a minor problem, such as a cold, could cost 500 RMB, almost one month's salary for some migrants. Financial problems also force some migrant workers to stop their treatment even after they are sent to hospital for emergency treatment. In 2002, fourteen migrant workers in a Beijing suitcase factory were sent to hospital by the local government when they were found to be suffering from severe benzene poisoning, but more than ten checked out soon afterwards due to lack of money (Xiao 2002). The Department of External Injuries in Guangdong Province People's Hospital receives approximately 200 migrant workers a year, and more than one-third cannot afford the treatment bill. Some hospitals now simply refuse to receive migrant patients (Cheng and Wen 2002).

Lack of access to financial help and proper treatment forces migrants to adopt some very unhealthy behaviours when becoming ill. Typically they adopt a 'wait and see' attitude in the beginning, hoping the illness will go away. If the situation gets worse, they go to small pharmacies to buy medicines according to their own medical knowledge. Only when the illness becomes unbearable do they visit hospitals, by which time the disease may have already become very serious. Gastric ulcers, for example, are common among migrants, but they often buy painkillers to ease the stomach-aches, meaning the gastric ulcer develops due to the delay in treatment. What is worse is that because migrants are young, in many cases they are able to initially endure the illness, but it may subsequently develop into a serious illness when they are older.

Lack of access to existing health services also severely impedes migrants from protecting themselves in certain circumstances. Migrant workers are usually very reluctant to take their employers to Court for a work-related illness, no matter how severe. Besides migrants' lack of legal knowledge, and lack of financial capacity to hire lawyers, a very important reason for this is that some employers pay for immediate medical treatment in order to save the migrant's life on condition that they promise not to sue the employer. Migrants simply cannot afford to shoulder the costs while waiting for the legal case to go through (for a typical case see Xiao 2002).

HIV and other STDs

HIV and other STDs have been the focus of recent research on migration and health. It was feared that migration may form a vicious cycle in the spread of STD viruses: migrants are more likely to be infected and, in turn, their mobility contributes to further spreading the virus, particularly by bringing the virus to the countryside where medical facilities are poorly equipped. There are also various reports pointing to the association between migration and STDs (see e.g. Shao 2002; Yang 2002, 2004). However, a causal relationship between migration and HIV and other STDs is very difficult to establish. The sex industry, for example, is often thought to be a link between migration and STDs. China's commercial sex industry has exploded in the last twenty years to include currently more than three million sex workers. HIV positive rates among sex workers tested in Guangxi and Yunnan in 2000 were 10.7 per cent and 4.6 per cent respectively (Kaufman and Jing 2002).

It is true that a large proportion of sex workers are migrants. Further, Pan Suimin (2002) estimated that sex workers typically stay in one place for only a few months. However, in many cases, migration may be a result, rather than a cause, of being involved in the sex industry. Sex workers have to move frequently in order to maintain anonymity and escape regular government campaigns to crack down on the industry. In other words, sex workers are better seen just as sex workers rather than as 'migrants', a category that also includes many different groups. Some scholars (e.g. Hansen and Li 2002; Yang 2002) speculate that migrants contribute to the spreading of virus as active patrons of commercial sex due to their separation from families and the local community. But a national survey conducted by the Renmin University Sexology Institute suggested that migrant workers make up only a minor part of commercial sex clients. On average, migrants buy sex more often than do rural residents, but less often than do urban dwellers. Enterprise managers and government officials are ten times more likely than male manual urban workers to patronise sex workers (Pan Suimin 2001, 2002), although migrant workers may be more vulnerable to STDs in individual cases, possibly due to their low usage of condoms (Hansen and Li 2002).

The statistical association between migration and STDs may also be a result of factors commonly affecting the two variables. Yang (2004), for example, cites the association between the numbers of temporary migrants per thousand permanent residents of a place, and that place's incidence of HIV and other STDs as evidence for the positive correlation between the two. However, it may well be the case that a place with more migrants is economically more prosperous and socially more open, and all the residents, both permanent and temporary, may be more prone to STDs. We should be particularly careful not to lend new excuses to xenophobia against migrants by overstating the linkage between migration and diseases.

The political economy of medical care

The health problems that migrants face have attracted some public attention, and researchers and activists have argued that the government must include migrant workers in a universal medical care system. While this should certainly be a long-term policy goal, my investigation suggests it to be unrealistic in the near future since the medical care system is subject to a series of more fundamental institutional arrangements. What follows first discusses why migrants are excluded in China's new medical care system at the national level, then, based on a review of local practices, explains why it is operationally unfeasible for the time being to include migrants in formal medical care. Finally, I identify the obstacles to the implementation of labour protection measures in the workplace – obstacles that are fundamentally attributable to local governments' lack of incentive to protect migrant workers, and migrants' lack of legal rights to establish their own organisations.

Rural–urban dualism and the political economy of medical care system reform

Migrants cannot be included in the formal medical care system first because of the rural–urban divide which has been a key characteristic of China's public policies (see Knight and Song 1999). In the 1950s, in order to prevent the rural population from spontaneously moving to cities, and to keep the grain price as low as possible to support rapid industralization in cities, particularly in heavy industries, the Chinese government established a special household registered system (the *hukou* system). Under this system, people born in rural areas could not move to the city and obtain urban *hukou* status unless mandated by the state (for literature on the *hukou* system, see Christiansen 1990; Cheng and Selden 1994; Mallee 1995; Chan and Li 1998). Although Chinese society has in many respects fundamentally changed since the end of the 1970s, the rural–urban dualism in social management largely remains the same. This makes internal migration in China an institutionally unique phenomenon in that migrants are not only undergoing change of residence and occupation, but are also being released from state control and the support system. Spontaneous migrants are generally not allowed to become urban residents, yet few are willing to return to rural areas. Therefore, they have become a special social category: the 'floating population' (see Solinger 1993; Xiang 1999).

Medical care is directly shaped by the rural–urban dualist structure. In the countryside, China used to have a well-established rural cooperative medical system that was set up at the end of the 1960s. Production Brigades diverted a proportion of their collective resources every year to form a fund to cover their members' medical costs.[3] At the same time, each Brigade had its own 'barefoot doctor' to conduct daily health inspections, and provide emergency treatment, and treatment for common illnesses. But this system collapsed in the 1980s when communes were dismantled and the new individual family-based Household Responsibility System was adopted. There is now little collective resourcing left to support the medical fund, and the management of the fund has also become a problem (Li Weiping 2002).

According to the 1998 National Health Services survey, the cooperative medical system covered only 1.83 per cent of farmers (Li Weiping 2002: 27). Furthermore, of all the investments in rural health care, the inputs from government dropped from 12.5 per cent in 1991 to 6.6 per cent in 2000, while the share shouldered by peasants increased from 80.7 per cent to 90.2 per cent (Li *et al.* 2003: 2–3). In theory, migrants are supposed to claim for medical care benefits in their home places because they are still registered there. But it is self-evident that given the very poor medical care system in rural China, migrants cannot expect anything from it. Apart from this, even in places with rural medical care, it is unrealistic to require migrants to go back to their home places for treatment, or to expect the rural authorities to reimburse the higher urban medical costs incurred.

The urban medical care system, being completely different from the rural system, faces a different crisis. Soon after the People's Republic of China was founded in 1949, China established a Public Fund Medical Care (*gongfei yiliao,* also translated as Government Employee Insurance Scheme) for government staff and those working for government-related units, and a Labour Security Medical Care (*laobao yiliao*, also translated as Labour Insurance Scheme) for state-owned or collective enterprise employees. Both offered completely free medical care, but the former was financed by the government, while the latter was financed by enterprises.[4] The two schemes share a unique feature in that both were tied up with work units (see Gu 2001). The work unit settled costs incurred by employees at the treating hospital, and there was no individual account, nor was there cross-work–unit coordination in managing the medical care fund (that is, risk pooling).

China started reforming its urban medical care systems in 1994 and a new nation-wide system was instituted in 1999. Under the new system, every employee has an individual medical account at a medical fund managed by local health security management bureaux. Both the employee and employer make monthly contributions to the account. When an employee needs medical treatment, the hospital contacts the fund directly to settle the bill, without involving the employer. Compared to the old system, the new scheme has three distinctive features, namely 'unification', 'sociali-sation' and 'individualisation'. 'Unification' means that the scheme requires all types of employees of all urban enterprises to join the system; 'socialisa-tion' means that the medical care fund is managed socially and not confined to particular enterprises; and 'individualisation' refers to the establishment of individual accounts, which enables a person to accumulate money in his or her account and to change jobs without changing the account.

However, despite the reform goal of 'unification', the new medical care scheme does not cover migrant workers. Documents issued by the Ministry of Labour and Social Security and the State Council refer to the new scheme as being for 'urban employees' (*chengzhen zhigong*).[5] Although the term 'urban employees' is not clearly defined, it is universally interpreted as referring to employees in urban enterprises with urban *hukou* status. When provinces such as Guangdong experimented in extending medical care to migrant workers, local governments had to issue separate documents. The reasons for this go beyond medical care itself and are related to more fundamental issues.

First, the immediate motivation for the central government to reform the medical care system was to relieve state-owned enterprises (SOEs) of the burden of shouldering almost unlimited medical care for their employees, which is believed to be a precondition for SOE profitability. However, including migrant workers in the medical care system does not to help achieve this goal – on the contrary, it may conflict with this objective. It has become common practice for SOEs to replace old employees who have generous welfare coverage with migrants lacking any fringe benefits. Apart from relieving SOEs of the burden, another key goal in reforming the medical care system, as well as other social security measures, is to maintain social and political stability. It is feared that certain groups may turn into 'elements of instability' if they are deprived of basic social security. However, compared to groups such as laid-off SOE workers and retired military personnel, who were among the most privileged in the pre-Reform era, but who lost most of their benefits in the process of marketization, migrants are not seen as a high-risk group in this sense. Related to the objective of ensuring social and political stability, the government is more willing to allocate resources for catastrophe relief than to offer basic medical care to all citizens universally, since the former is more effective for achieving that goal.

Second, the vested interests associated with the old welfare system form another obstacle to including migrants in the formal medical care system. As mentioned above, the old medical care system is linked directly to work units, and different types of work units offer very different medical care. Government departments and large SOEs often designate well-equipped hospitals to treat their staff and cover all the costs, while staff of collective enterprises can only go to small hospitals with limited subsidies. Although the new scheme claims to unify this, the vested interests are so deeply ingrained that in practice the boundaries between different types of work units have hardly been challenged. Persuading groups with vested interests to make sacrifices and join a unified medical care system is difficult enough for the government, and no party is willing to represent migrant workers' interests in the bargaining.

Third, there are concerns that migrants will create extra difficulties for the management of the medical care system. It is feared, for example, that the rural population may rush to the city when sick and falsely claim to be migrant workers in order to benefit from the medical care, or migrants may use their insurance to cover their family members' treatment costs.

Conflicts between formality and informality, between mobility and locality

The above section clarified how including migrant workers in China's formal medical system may go against the government's immediate aim of medical care reform. What follows identifies the operational obstacles to including migrants in the system even when the authorities are willing to do so. Since the late 1990s, places such as Shenzhen, Zhuhai and Dongguan, all cities in Guangdong province with large numbers of migrant workers, have mandated that all enterprises should offer medical insurance to migrant workers. But the results are far from satisfactory. Shenzhen has 3.3 million registered migrant workers – the actual size may be five million – but by the end of 2001, approximately 1.04 million persons, including both permanent residents and migrants, had medical insurance (Cheng and Wen 2002). Given that most permanent residents join the medical care system, only very few out of the 1.04 million would be migrant workers.

The operational difficulty of migrants being included in medical care stems first from the fact that many migrants enter the job market informally, while the medical care system is strictly employment-based in needs operation.[6] It is the employer who sends the money to the medical fund, partly deducted from the employee's salary and partly from the employer. Although labour bureaux conduct inspections of enterprises from time to time to ensure that all the workers have been insured, given the oversupply of labour and, in particular, the very disadvantaged position of migrants in the job market, migrants commonly work without any formal contract. If the worker insists on having a contract, the enterprise may reduce the wage as a precondition on the grounds that a formal contract would cost the enterprise more by obliging it to comply with government regulations and pay various levies. A 2002 inspection conducted by the Labour Bureau of Chang'an town, Dongguan city, Guangdong province, found that a large proportion of private enterprises grossly under-reported the numbers of their employees. A toy factory hired approximately 1,000 workers, but paid medical insurance for only 150. According to a survey conducted by Tan and her associates in the Pearl River Delta in 1994, only 10 per cent of the workers had a formal contract with the employer and approximately 60 per cent had never signed an employment contract (Tan 2002c: 312). In many cases where migrants sue employers for mistreatment, the biggest problem turns out to be that the migrant cannot even prove that he or she was employed. Apart from that, a large proportion of migrants in large cities are self-employed. The 1999 consensus of migrants in Beijing, for example, showed that more than 40 per cent of all the migrants were self-employed (Liu and Wu 1999), did not have a 'work unit' to belong to, and therefore could not join the medical care scheme.

It must be recognised that informal employment, namely employment relationships without legally effective contracts, have existed for a long time in China. In the coming ten to fifteen years, informal employment is projected by some scholars to contribute half of the total urban employment opportunities (Hu An'gang (2001), cited in Baozhang Ketizu 2002: 2). Sometimes migrants themselves prefer informal employment relationships and do not want to sign contracts. This is because the option of quitting a job immediately is often migrants' only and last resort when the situation in a factory becomes unbearable. A formal bond would deprive them of even this opportunity (Xiang 1995; Tan 2002b: 67). It is not advisable either for the government to force all enterprises to adopt formal employment relationships with social welfare benefits. The World Bank has pointed out in its 1990 *World Development Report* that undue regulations in the labour market, including compulsory social security, increase labour costs, and therefore reduce the demand for labour. As a result, stricter labour market regulations benefit those in the advanced employment sector, but worsen the unemployment problem for the poor. Undoubtedly, employment should take precedence over social security in China today (see Baozhang Ketizu 2002: 3).

The conflict between migrants' extraordinarily frequent mobility and the localised operational pattern of medical care funds also makes it difficult to include migrants in the formal medical care system. The scope of risk pooling ('social coordination' as it is usually referred to in China) of the medical fund is now at county level (or small cities or districts in large cities, which are equivalent to counties in administrative rank), and a nation-wide unified system is far from being established. A worker's medical care account has to be cancelled from one health security management bureau and reinstated in another if he or she changes jobs across counties. It is even more troublesome if the worker moves from one province to another since there is almost no connection or coordination across provinces in terms of medical care. If an urban migrant worker returns to the countryside, as they often do, then the medical insurance has to be cancelled altogether. The lack of any effective medical care system in rural areas makes it literally impossible to transfer one's account from an urban centre to a rural area. Because of the tension between migrants' high level of mobility and the accounts' low portability, even in cases where employers have registered their migrant workers for medical care, workers often cancel their accounts later, particularly around Chinese New Year when many of them change jobs or locations. For example, 17,817 people withdrew their medical care accounts in Chang'an town, Dongguan city, in the first three quarters of 2001, most of whom were presumably migrants. In 2001, 120,000 migrant workers withdrew their pension insurance in the same town (Cheng and Wen 2002).

The relationship between government, employer and migrant worker

As demonstrated above, work-related illness constitutes a major health threat to migrant workers, but in fact there are a series of relevant regulations in place. In October 2001, for example, the National People's Congress passed the Law of the People's Republic of China on the Prevention and Treatment of Occupational Diseases, which clearly stipulates that workers must be fully informed of the dangers related to work, employers must take adequate preventive measures, and employers should provide treatment in cases were illness occurs. The law also designates that employers' obligations should be clearly indicated in labour contracts. Some provinces and cities, such as Beijing, also require enterprises with a high risk of work-related accidents to pay special insurance for their employees' work-related illnesses. Other regulations specifically aim to protect female workers' health, including the Regulations on Labour Protection of Female Workers (1988), the Regulations on Activities not Suitable for Female Workers (1990), and the Regulation on Female Workers' Health Protection (1993).

But these laws and regulations fail to protect migrant workers, particularly those in Township and Village Enterprises (TVEs) and small private enterprises, who are the major employers of migrant workers. A news story in the *China Woman's News* reported that private and foreign-invested shoe factories in Putian, Fujian province, earned billions of RMB since 1984 by using female migrant workers, but were not willing to divert even 1 per cent of their profits to improving the working conditions that had destroyed the health of thousands of workers (*China Woman's News* 6 January 1996, cited in Tan 2002e: 241).

The failure of formal regulations is, to a large extent, due to the unbalanced power relationship between the government, employers and migrant workers. It must be borne in mind that in China government intervention remains far more powerful than legal intervention in terms of regulating social and economic life. Local governments are generally reluctant to intervene to protect migrant workers' rights against enterprises' interests. While the traditional SOEs were built as 'working-class paradises' rather than as profit-making firms, TVEs and private enterprises run to the other extreme, and profit-making takes precedence over almost everything else. For local governments, TVEs and other private enterprises are not only important fiscal sources; the revenue is also taken by the higher level of authorities as a key criterion in evaluating local government's performance. In 1993, the government of Kuiyong Town under the Shenzhen Municipality wrote to the Shenzhen Fire Prevention Enforcement Team to urge them to issue a fire safety certificate to a Hong Kong businessperson who had invested in the Zhili Toy Factory in Kuiyong Town. The letter threatened that 'if [the certificate] is not given, the economic development of Kuiyong Town will be affected, and Hong Kong investors will organise a collective complaint to the municipal government directly' (*Workers' Daily* 26 December 1993, cited in Tan 2002b: 69). Soon after the letter was sent and the certificate was issued, a fire broke out in the factory and killed eighty-seven migrant workers.

What is more fundamental is the relationship between the employers and the workers themselves where migrant workers have literally no bargaining power. More than 95 per cent of workers in non-state-owned enterprises are not unionised (Weldon 2001/2002: 28). Even for those enterprises which have unions, the union is very different from what it is supposed to be internationally. According to orthodox Marxist–Leninist theory, workers in a socialist state are the owners of the state and are fully represented by the Party, and there is no labour–capital struggle anymore. Therefore, trade unions in China are mainly concerned with issues such as entertainment, education and other welfare (such as purchasing food collectively at a lower price for the workers). Furthermore, all trade unions in China must be branches of the All-China Federation of Trade Unions (ACFTU), and the ACFTU works strictly under the Party's leadership.

The highly skewed relationship between migrant workers and employers in private enterprises is not merely a reflection of the general antagonism between labour and capital; it is compounded by the role of local government. Under the current *hukou* system, government departments have no responsibility for those who are not formally registered with them. In many towns along China's eastern coast and particularly in the Pearl River Delta, for example, migrants have significantly outnumbered the local population, but these migrants are very rarely mentioned in local government's development plans and reports. All the social and economic development indictors, such as per capita income and numbers of hospitals per thousand people, are calculated on the basis of the size of the permanent population. Therefore, in a sense the existence of cheap, unprotected migrant labour is in the local government's interest since this labour will attract foreign investment, boost land rent rates, and dramatically increase the income levels of local people (for a brief discussion on the development model based on the alliance between foreign investors and local government on the one hand, and migrant workers' cheap labour on the other, see Xiang 1995).

Conclusion

This chapter has offered an overview of the health risks that rural–urban migrants in China face and the basic policy gaps in providing them with access to basic health services. Existing literature on migration and health, be it from a medical or a sociological perspective, tends to attribute migrants' health problems to their behaviour patterns and social characteristics, such as supposedly active premarital and extramarital sexual activities, high levels of mobility, low incomes, lack of awareness, and lack of social contact with the local community (see e.g. Yang 2002). All these arguments are valid in China's case, but more importantly, as this chapter has demonstrated, migrants' health problems are a result of existing formal institutional arrangements. The fact that migrants face high health risks at work is seemingly a result of market segmentation, but fundamentally it is attributable to the unbalanced relationship between migrants, enterprises and local government, which is, in turn, related to the *hukou* system and the current growth regime in China.

The existing institutional arrangements not only render migrants vulnerable, but also impede them from being included in the formal medical care system. More specifically, the reasons include: vested interests and the government's immediate goal in reforming the social security system which provide the state with no motivation to do so; the gap between the urban and the rural in the medical care system which makes it impractical to provide migrants with formal medical care; the localised, and therefore geographically fragmented, operational pattern which also discourages migrants from joining the system; and finally, the informal employment relationship prevalent among migrant workers which conflicts with the medical care system's reliance on formal employment contracts for implementation.

Based on this assessment, we may need to move the focus away from the medical care system itself and place more emphasis on the overall institutional set-up as well as something more attainable in the short term. Since changes in institutional arrangements will take a long time, grass-roots-level activities that can be initiated in a relatively short period of time are needed to address migrants' immediate health problems. For instance, my fieldwork suggests that community-based health services provided by community clinics are an effective measure. Other activities that the government may consider include health information campaigns, allowing for, or even encouraging, the establishment of clinics by migrants themselves, empowering migrants by providing legal assistance, and developing migrant self-help organisations. In the final analysis, migrants in China face special health problems not only because they are a new social group, but more importantly because they cannot be incorporated into the state system, yet at the same time the state remains the single provider of social welfare. The Chinese government must realise that it is unrealistic to cover all its citizens' welfare without cooperating with other social forces, and accordingly that an active civil society may be indispensable to maintaining the stability of an increasingly diversified society.

Notes

1 I would like to thank the Asian MetaCentre for Population and Sustainable Development Analysis and the Asia Research Institute, National University of Singapore, for their support of this research. I would also like to thank Tan Shen, Shi Xiuying, Zhe Xiaoye, Tang Jun, Yang Tuan, Huang Ping, Xiong Pingyao, Yuan Yue, Gong Sen, Li Weiping and Yang Xiushi for their assitance in my investigation. Theresa Wong, Gu Xin and Doreen Montag offered very helpful suggestions for revision of earlier versions of the chapter.

2 Construction forms a major occupation in which migrant workers are concentrated. Approximately one-third of all economic migrants (including self-employed) in Beijing in 1999, for example, were construction workers (Liu and Wu 1999).
3 A Production Brigade is the equivalent of an Administrative Village after the Reform. Typically one Production Brigade comprises 1,000 to 5,000 people.
4 See the Political Department (now State Council), *Regulations of the People's Republic of China on Labour Security* (1951, Zhonghua renmin gongheguo laodong baoxian tiaoli); Political Department, *Directives on Implementing Public Fund for Medical Treatment and Prevention for State Staff of All Levels of Government, Parties, Organisations and the Work Units Belonging to Them* (1952, Guanyu quan'guo geji renmin zhengfu, dangpai, tuanti ji suoshu shiye danwei de guojia gongzuo renyuan shixing gongfei yiliao yufang de zhishi).
5 See e.g. State Council, *Suggestions on Experiment Spots on the Reform of Employees' Medical Care System* (1995, Guangyu zhigong yiliao zhidu gaige de shidian yijian); State Institution Reform Committee *et al.*, *Suggestions on Expanding of the Experiment Spots on the Reform of Employees' Medical Care System* (1996, Guangyu zhigong yiliao zhidu gaige kuoda shidian de yijian).
6 I would like to thank Dr Sara Cook from the Ford Foundation in Beijing for helping me to clarify this idea.

References

Baozhang Ketizu (China Social Security System Reform Research Team, Yiwu Investigation Group) (2002) *'Chengshi pingkun de shenceng yuanyin: jiuye yu baozhang de erlv beifan'* (The root cause of urban poverty: the paradox between employment and social security), *Shehui Zhengce PingLun* (Social Policy Review), April: 1–3.

Chan, K.W. and Li Zhang (1998) 'The Hukou System and rural–urban migration in China: processes and changes'. The Center for Studies in Demography and Ecology Working Paper, University of Washington.

Chen Yonghui (1996) *'Mazu you lei'* (Tears of Mazu), *'Chengmo de gaoyang'* (The silent lamb), *'Shiheng de tianping'* (Unbalanced scale), *China Woman's News*, 15–17 January.

Cheng Yifeng and Wen Yuanzhu (2002) *'Shui lai wei tamen tigong baozhang? Nongminggong shebao wenti yinren guangzhu'* (Who will provide security for them? Migrant workers' social security deserves attention). Xinhua News Agency, 21 June.

Cheng, T.J. and M. Selden (1994) 'The origins and social consequences of China's *hukou* system', *China Quarterly*, 139: 644–68.

Christiansen, F. (1990) 'Social division and peasant mobility in mainland China: the implications of *hukou* system', *Issues and Studies*, 26 (4): 78–91.

Dai Shulin, Meng Pengtong and Zhong Xuan (2001) *'Taishun xi feng an yishen puanjue'* (First trial concluded on quartz lung case in Taishun), *Wenzhou Shangbao* (Wenzhou Commercial News), 25 October.

Gao Yu (2001) *'Fei qian gei hui le'* (Lungs completely destroyed), *Beijing Qingnian Bao* (Beijing Youth News), 7 November.

Gu, E. (2001) 'Market transition and the transformation of the health care system in urban China', *Policy Studies*, 22 (3/4): 197–215.

Guan Xingping and Jiang Miaoyi (2002) *'Chengshi wailai renkou de jiben shenghuo yu jiankang fuwu*' (Basic life and health services for migrants in cities), in Li Peilin (ed.) *Nongmingong, Zhongguo Jincheng Nongmingong de Jingji Shehui Fenxi (Peasant Workers: Economic and Social Analysis of Peasant Workers in the City)*, *Shehui Kexue Wenxian Chubanshe* (Social Sciences Documentation Publishing House).

Hansen, P. and Li Meng (2002) 'Sexually transmitted infection risk in migrant construction worker populations in Beijing, China', Paper presented at the Annual Meeting of the Association of American Geographers, 22 March.

Huang Ping and Frank Pieke (2003) *China Migration Country Study*, A Report for the UK Department for International Development, 1 May.

Kaufman, J. and Jing Jun (2002) 'China and AIDS – the time to act is now', *Science*, 296: 2339–40.

Knight, J. and Song, L.N. (1999) *The Rural–Urban Divide: Economic Disparities and Interactions in China*, Oxford: Oxford University Press.

Li Weiping (2002) *Zhongguo Nongcun Jiankang Baozhang de Xuanze* (Options for Rural Health Security in China's New Policy Environment Synthesis Report), Beijing: *Zhonggou Caizheng Jingji Chubanshe* (Chinese Fiscal Economy Press).

Li Weiping, Shi Guang and Zhan Kun (2003) *Woguo nongcun weisheng baojian de lishi, xianzhuang yu wenti*' (The history, current situation and problems in China's rural health care). Unpublished paper.

Li Weisheng (2002) *Jiaqiang liudongrenkou guanli, jianshao mianyi kongbai renqun*' (Strengthening the management of the floating population, downsizing the vacant groups for vaccination), *Weisheng yu Jiankang* (Hygiene and Health). Online, available at: www.cdcp.org.cn/newgc/jiaqiang.htm (accessed 2 July 2003).

Liu Shuang and Wu Xiaoping (1999) *Dui dachengshi wailai laodongli liuru de sikao*' (Thoughts on migrant labour in large cities), *Zhongguo Renkou Kexue* (Chinese Population Science), 3: 46–52.

Mallee, H. (1995) 'China's household registration system under reform', *Development and Change*, 26: 1–29.

Pan Guanglin, Tan Kelong and Li Jianping (2001) *Quanguo shouli jiti "xi fei an" gongshang suopei an jishi*' (A report on the first legal case for seeking compensation for occupational 'quartz lung disease' work in China), *Fazhi Ribao* (Law Daily), 8 June.

Pan Suimin (2001) *Aizibing zai Zhongguo: Xing chuanbo de kenengxing jiujing you dou da?*' (AIDS in China: how likely it is transmitted through sexual channel?). Online, available at: *Sohu Net* health.sohu.com/87/44/harticle15534487.shtml (accessed 2 November 2001).

Pan Suimin (2002) 'Rough trade, rough justice', *China Development Brief*, 5, 1: 36–8.

Shao Qing (2002) *Nvxing xidu zhe yu Aizibing*' (Female drug takers and AIDS), *Falu yu Shenghuo* (Law and Society), 12.

Solinger, D.J. (1993) 'China's transients and the state: a form of civil society?' *Polities & Society*, 21, 1: 91–122.

Tan Shen (2002a) *Waichu he huixiang: gei nongcun liudong nvxing dailai le sheme?*' (Going out and returning: what do they bring to rural women?), *Funv yu Laogong* (Women and Labour), Unpublished collection, pp. 139–48.

Tan Shen (2002b) *Jujiao zhujiang sanjiaozhou: wailai nvgong yu waizi qiye, dangdi shehui zhi guanxi*' (Focusing on the Pearl River Delta: the relationship between female migrant workers, foreign-invested enterprises and the local government), *Funv yu Laogong* (Women and Labour), Unpublished collection, pp. 61–73.

Tan Shen (2002c) *1990–1995 zhongguo fuvu fazhan baogao*' (Report on Chinese women's development, 1990–1995), *Funv yu Laogong* (Women and Labour), Unpublished collection, pp. 298–323.

Tan Shen (2002d) '*Zhujiang sanjiaozhou wailaigong diaocha fenxi*' (An investigation and analysis of migrant workers in the Pearl River delta), *Funv yu Laogong* (Women and Labour), Unpublished collection, pp. 74–84.

Tan Shen (2002e) '*Zhujiang sanjiaozhuo wailai dagongmei de xianzhuang yu fazhan*' (The current situation and development of the situation of the female migrant workers in the Pearl River Delta), *Funv yu Laogong* (Women and Labour), Unpublished collection, pp. 235–51.

Weldon, J. (2001/2002) 'New moves for Pearl Delta migrants', *China Development Brief*, 4, 3: 25–7.

Wen Tiejun *et al.* (2003) '*Guoren Xiangcun Jianshe Shiyanyuan Jianyishu*' (Proposal for the reconstruction of rural schools). Unpublished paper.

Xiang Biao (1995) *Minggong Wenti yu Chengshi Shehui Chengzhang* (Peasant Workers and Urban Development: A Case Study of DongGuan City, Pearl River Delta), The Department of Sociology, Beijing University.

Xiang Biao (1999) 'Zhejiang Village in Beijing: creating a visible non-state space through migration and marketized traditional networks', in Frank Pieke and Hein Mallee, (eds) *Internal and International Migration: Chinese perspectives*, Richmond, Surrey: Curzon Press, pp. 215–50.

Xiao Xianwu (1999) '*Wailai liudong renkou ertong baojian xianzhuang diaochao*' (An investigation of the current situation of migrant children's health protection), *Zhongguo Fuyou Baojian* (Women and Children's Health Protection), pp. 761–2.

Xiao Yi (2002) '*Yi nian sunshi 4.5 yi, 'zhiyebing' rang Beijing tong xia shashou*' (450 million lost a year, occupational diseases force Beijing to act). Online, available at: www.21dnn.com.cn (accessed 11 April 2002).

Yang Xiushi (2002) 'Migration, socioeconomic milieu, and HIV/AIDS: a theoretical framework'. Unpublished paper.

Yang Xiushi (2004) 'Temporary migration and the spread of STDs/HIV in China: is there a link?', *International Migration Review*, 38, 1: 212–35.

Yu Xiaomin (2001) '*Kuayue jieji de bianjie*' (Transcending the class boundaries). Paper presented at the International Conference on the Mobility of Rural Labour Force, 3–5 July, Beijing.

Zhang Zhen (1999) '*Liudong renkou zhong yulin funv shengzhi jiankang de xianzhuang yu sikao*' (The current reproductive health situation of female migrants of reproductive age and some thoughts), *Renkou Xuekan* (Population Journal), 4: 55–9.

10 A longitudinal analysis of health and mortality in a migrant-sending region of Bangladesh

Randall S. Kuhn

Introduction

Recent theoretical and methodological advances in the study of migration have expanded analysis of its impact from the well-being of migrants to include the well-being of family members left behind in the household or village of origin. The importance of migration, both as loss of local support and labour, and as a means of diversifying and enhancing a household's livelihood, provokes interest in the migration/health relationship, particularly for the elderly, in rural regions throughout less developed countries (LDCs). Yet thus far little work has tried to measure the impact of migration on the health or survival of the left behind family.

In Matlab, a rural district of Bangladesh, rates of out-migration to cities and abroad have grown even as overall family sizes have begun to decline. As a result, an increasing proportion of elders can expect a substantial number of their children to live away from home. Therefore, it is important to understand whether parents do indeed benefit from the migration process and, if so, whether the benefit extends to better health. New survey designs have made it easier to approach these issues, by facilitating micro-level linkages between household survey data on socioeconomic status, kinship and mobility to self-reported and objective measures of health, and to longitudinal data on health and mortality (Rahman and Barsky 2003; Frankenberg and Thomas 2000).

This chapter examines the impact of children's migration on the health and survival of older respondents to the 1996 Matlab Health and Socioeconomic Survey (MHSS), a multi-stage, multi-sample survey of household economics, health and social networks. Matlab is the site of an ongoing Health and Demographic Surveillance System (HDSS), maintained and collected by the International Centre for Health and Population Research (known as ICDDR, B). When matched, these two data sources permit analysis of mortality in the years subsequent to the collection of detailed household survey data. In particular, the study considers the impact of migration on health and mortality; the impact of internal and international moves and moves by sons and daughters; the effects of migrant education; and the extent to which migrant financial transfers, or remittances, explain any positive effects of migration on health.

Theoretical background

Studies of the health of left-behind populations represent the inevitable intersection between two literatures: on social networks and health, and on migration and transnational networks. Both literatures address processes of modernization that were once thought to diminish family ties, the first between parents and children, the second between migrants and non-migrants (Boserup 1965). Both literatures find such expectations to be overly simplistic: families often facilitate and invest in processes of modernization and change such as migration, and those that do not participate directly in these processes often draw considerable benefit from continued social and economic cooperation with those who do (Coleman 1993; Stark 1982).

Social networks and mortality

Modernization theories of the family addressed the effects of increasing urbanization, education and labour market participation stemming from the shift from traditional to modern agrarian modes of production (Boserup 1965). Changing roles, declining parent–child contact and an associated rise in children's bargaining power with parents was expected to lead to a crisis of support for the aged, such as that experienced in the West during the Great Depression (Cowgill 1986; Parsons 1943). Yet recent fertility and urban transitions, particularly in Asia, provide little evidence of a negative relationship between demographic change and old-age support (Asis *et al.* 1995; Hermalin and Myers 2002; Kabir *et al.* 2002; Mason 1992). Many of the burdens of demographic change are, in fact, borne not by parents facing fewer and more distant support options, but by children, who must share support obligations among fewer siblings or who must live apart from spouses and children in order to maximize parental well-being (Knodel *et al.* 1992; Kuhn 2001). Studies of old-age health suggest vulnerability to demographic change among marginalized populations such as women, widows and the very elderly, yet studies are more likely to find a positive relationship between modernization and old-age health, both across settings and over time (Iwasaki *et al.* 2002; Knodel *et al.* 1992; Rahman 1999; Rahman *et al.* 1992; Wu and Rudkin 2000).

While some have argued that the tendency towards consistent support from migrant children results from stronger family values in settings such as Asia as opposed to the West, social network theories instead address how families in some settings directly encourage and invest in individual opportunities such as education and migration in order to generate mutual benefit and security (Coleman 1988; Litwak 1960; Portes and Sensenbrenner 1993). The family plays a particularly important role when government or market-based institutions for credit, investments or retirement insurance remain underdeveloped. Parents in industrializing societies favour a small number of well-educated, mobile children over large families due to the high returns and low risks to such investments (Becker and Tomes 1976; Lillard and Willis 1994).

Since families themselves guide processes of modernization and change, social network theories do suggest that old-age support in such periods of social change will continue to reflect a society's pre-existing kinship norms and preferences (Asis *et al.* 1995; Skinner 1997). A study of intergenerational transfers in historically patriarchal Taiwan, for example, found that sons persisted as the paramount source of old-age support in spite of rapid increases in women's wages and labour market participation (Lee *et al.* 1994). Research in Bangladesh has suggested that sons play a greater role in predicting old-age health and survival than do daughters, although more recent work suggests a narrowing of such gender preferences over time (Kabir *et al.* 2002; Rahman 1999; Rahman *et al.* 2004).

Migration and health

Research on migration has similarly moved beyond the modernization hypotheses, which suggested that migration, as an individual wage-maximization decision, would only affect stayers' well-being through decreased household labour supply (negative) and decreased competition for resources and opportunities (positive) (Harris and Todaro 1970). Two distinct literatures have extended theories of social networks and the family to incorporate migration. The 'New Economics of Labour Migration' suggests that households in LDCs use migration to ease liquidity constraints, or limited access to capital markets, by diversifying family livelihoods and increasing access to investment capital for both movers and stayers (Lucas and Stark 1985; Massey *et al.* 1999). Theories of transnationalism relate international migration to broader transnational networks of social, economic and political solidarity that typically emerge from migration processes, but may also include cross-border flows of investment, trade and travel oriented around ethnicity or community (Faist 2000; Portes 1996).

These theories both emphasize social and economic connections between migrants and those left behind. The New Economics suggests that migrants may not necessarily move to areas having the highest wages, but to areas that offer short-term economic opportunities which maximize the needs of both movers and stayers (Stark and Bloom 1985). Migrants often minimize destination-area consumption in order to remit earnings back to their place of origin (Massey and Parrado 1994). Transnational theories address the tendency for migrants to choose the most socially expedient destinations, including those with high concentrations of past migrants from the same community and those that favour quick earning opportunities over long-term settlement opportunities (Massey and Espinosa 1997). Exchange between movers and stayers, whether manifest by a simple flow of remittance income or a complex transnational network of capital and information flows, has a measurable and typically positive effect on the economic and social well-being of those left behind, and is encouraged by stayers' investments in migration costs (Durand *et al.* 1996; Massey *et al.* 1999).

These theories have been applied to a growing literature, well represented in this volume, demonstrating the poor health status of some international migrants (see e.g. Chapter 1, this volume). Efforts to minimize consumption in the destination society may sacrifice current health care opportunities for future savings and investment opportunities. Furthermore, migration to nations that actively restrict guest workers' rights to health care and/or permanent settlement may limit migrants' freedom of movement and access to health care, and encourage further reductions in destination-society consumption. Stayers may thus be likely to draw substantial short-term health benefits from movers' sacrifices, particularly through consumption of migrant financial transfers, yet little research has actually addressed the impact of migration on the health or survival of original family members (Frank and Hummer 2002; Kanaiaupuni and Donato 1999). Existing evidence relating social networks to health suggest a strong possibility that migration, particularly by those children who would have been likely providers of support in the absence of migration, could have a substantial positive impact on parental health. Furthermore, parents may derive particular health benefits if they made more substantial investments in migration opportunities or education, or if migrant children went to destinations in which their likelihood of permanent settlement is particularly low.

Migration, urbanization and old-age support in Bangladesh

Matlab, even more so than other areas of rural Bangladesh, has an advanced process of migration both to Bangladeshi cities and abroad. It overlaps with ongoing processes of mortality and fertility decline, fuelling growing concerns over the health of the aged.

Migration and urbanization in Bangladesh

Although it remains one of the least urbanized countries in Asia, Bangladesh has urbanized rapidly since gaining independence in 1971. The proportion of Bangladesh's population living in cities (just 8 per cent in 1971) increased to 25 per cent in 2000, as the population of Dhaka, the capital city, increased from 1.3 million to over nine million (United Nations 2003). Medium-level United Nations projects estimate that the proportion of Bangladeshis living in cities will rise to 38 per cent by 2020, or higher than that of several wealthier Asian nations (United Nations 2003).

Throughout the 1980s and 1990s, international migration also grew in both demographic and fiscal importance. In 2000, one million overseas workers (0.7 per cent of the total population) sent a total of US$1.5 billion (3.9 per cent of total gross domestic product) to Bangladesh (United Nations Population Division 2002). In the late 1990s exports of human resources were the second or third largest source of foreign currency earnings in Bangladesh. Bangladeshis who wish to work overseas tend to migrate under contract, for fixed periods of time, to destinations in the Middle East or Southeast Asia, typically working in unskilled service or construction positions (Shah and Menon 1999). Migrants typically acquire few skills while overseas, and return to Bangladesh at the end of their contract.

Migration and Urbanization in Matlab

Matlab experiences intense migration flows. It is just six hours or less by launch from Dhaka and the industrial belt of southeastern Bangladesh; its relative proximity to major destination areas reduces the cost of migration and promotes the formation of social networks encouraging further migration (Kuhn 2003). Figure 10.1 shows the extent of men's internal and

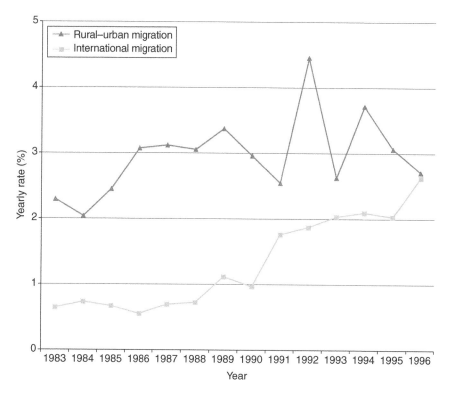

Figure 10.1 Trends in international and rural–urban migration, Matlab DSS Area, 1982 to 1996 (source: HDSS, 1982–1996).

international migration from the Matlab study area, depicting age-standardized out-migration rates from the Matlab HDSS area between 1982 and 1996. Internal out-migration rates for males aged fifteen to fifty typically stood above 3 per cent per year, an exceptionally high rate by global standards (Rogers and Castro 1981). International out-migration rates, also quite high by global and Bangladeshi standards, rose rapidly throughout the 1990s, peaking in 1996 at 2.7 per cent, only slightly lower than that year's internal migration rate. Between 1992 and 1996 in the Matlab HDSS area, 47 per cent of international migrant men in the fifteen to fifty age group moved to destinations classified as 'elsewhere in Asia', 42 per cent moved to the Middle East, and 10 per cent moved to India, leaving less than 1 per cent who moved to North America, Europe, Australia and all other countries combined.

The practices of solo migration and circular migration are common among international and internal migrants, even after marriage. Limited wages and benefits, as well as high urban housing and childrearing costs, encourage internal migrants to take advantage of low travel costs by shuttling between urban and rural areas (Afsar 1999). The pattern of return migration in Matlab reflects not only the tendency for migrants to fail and quickly return home, but also the gradual return of many migrants after years in the destination (Kuhn 1999). High travel costs necessitate solo moves for international migration, while restrictions on permanent settlement encourage return migration. The extent of solo male migration may be seen in male–female sex ratios from the 1996 Matlab HDSS census, which stand at 1.08 for the ten to fourteen and fifteen to nineteen age groups, but drop to 0.95 for ages twenty to twenty-four, 0.78 for twenty-five to twenty-nine, and 0.79 for thirty to thirty-four (Mostafa *et al.* 1998).

The importance of sons' migration largely reflects patrilineal kinship norms favouring old-age support from sons and their wives, and exogamous marriage by daughters (Frankenberg and Kuhn 2004). Sons in Bangladesh have been described as the best risk insurance available in old age (Arthur and McNicoll 1978; Cain *et al.* 1979). While recent work shows a narrowing of male–female differences in education and communication after marriage (Kuhn and Menken 2002; Simmons 1996), analysis of financial transfers from the data used in this study shows that daughters played virtually no role in intergenerational support networks (Frankenberg and Kuhn 2004).

Health and mortality transitions

Bangladesh has also seen improvements in health and survival through the post-liberation period. Rapid improvements in child mortality have been accompanied by more gradual advances in adult mortality due to improved income, nutrition, health care and maternal care. Figure 10.2 charts a gradual decline in age-standardized mortality rates for adults aged fifteen to fifty between 1982 and 1996. Bangladesh is also beginning to experience a process of population ageing and increasing old-age dependency. Total fertility rates have fallen from 6.5 in 1979 to 3.3 in 2000 (NIPORT, MA and ORCM 2001), and declining fertility and adult mortality have led to steady growth in the population over age fifty relative to those age fifteen to fifty, also shown in Figure 10.2. This trend towards increasing old-age dependency will accelerate as cohorts with higher adult survival rates move into old age, and smaller cohorts born since 1979 grow older (Kabir *et al.* 2002). The tendency for young adults to leave rural areas and return in old age, as discussed above, will further raise the ratio of rural elderly expecting support from adult migrant children.

Data and methods

Data for the study result from the integration of Matlab survey and demographic surveillance datasets. Cross-sectional health status and control measures come from the Matlab Health and Socio-Economic Survey (MHSS), which surveyed 11,200 individuals aged fifteen and over in 4,538 households. While designed for comparability to similar nationally representative family life surveys such as the Indonesian and Malaysian Family Life Surveys (IFLS, MFLS), MHSS eschewed a nationally representative sample in favour

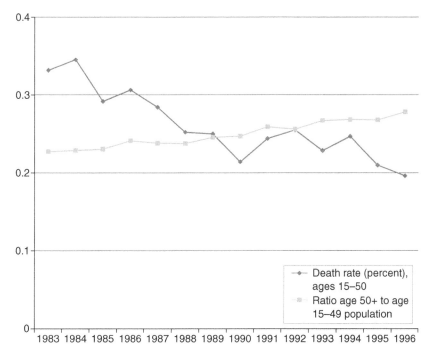

Figure 10.2 Trends in adult mortality and dependency ratios, Matlab DSS Area, 1982 to 1996 (source: HDSS, 1982–96).

of a sample based entirely in Matlab, where the ICDDR, B has operated the Matlab Health and Demographic Surveillance System (HDSS) since 1966 (Rahman *et al.* 1999). Matlab HDSS data have been used extensively in the demographic literature and are considered to be one of the few high-quality (that is, complete, accurate and up-to-date) data sources in the developing world (Faveau 1994). In particular, age reporting is considered to be highly accurate, a feature not found in other South Asian data sources (Menken and Phillips 1990).

The HDSS constitutes the second data source for this study. As part of a collaborative study between the Harvard School of Public Health, the University of Colorado at Boulder and the ICDDR, B, MHSS respondents were tracked through HDSS from 1996 through 2000. The resulting database identifies respondent death and censoring in each calendar year of the study. Linking survey and surveillance data in this way offers a number of advantages over panel survey data (Hummer *et al.* 1998). Foremost among these is cost, since the approach requires only one round of survey data. It also offers more reliable data on the timing of mortality or censoring, and better distinguishes between the two. Panel surveys in LDCs typically faced attrition rates of 10 per cent or more, resulting in bias if death and migration cannot be properly identified as the cause of exit, and survey costs can rise considerably with intensified follow-up (Frankenberg and Thomas 2000).

The analysis focuses on 1,666 women and 1,719 men aged fifty and above (3,385 total) who responded to the MHSS individual survey module. The MHSS provides data on the respondent's health status at the time of survey based on general health self-reports, self-reported physical disability and objective tests of physical function (see below for greater detail). Analysis is restricted to the random subsample of respondents who performed physical tests, comprising 1,475 women and 1,581 men (3,056 in total). These data are used to generate dependent variables in models of health status at the time of the survey, as well as crucial statistical controls in longitudinal mortality models.

Models include statistical controls for the respondent's own age, sex, marital status, educational attainment and employment status (not shown in the tables). By matching spouses to one another, their age, education, employment status and co-residence are also included. Detailed household economic modules provide controls for household landholdings and income, which are crucial given the strong relationship between household assets and migration, particularly to overseas destinations (Kuhn 1999). Observations are weighted to reflect the true probability of household selection (Rahman *et al.* 1999).

Cross-sectional health status models

Cross-sectional models were constructed to predict the probability that a respondent is in good health at the time of survey in terms of children's characteristics, with controls for respondent/spouse characteristics and household wealth. A set of dichotomous variables (1 if the respondent has severe disability, 0 if the respondent is in relatively good health) was constructed along three dimensions of health: overall self-reported health, self-reports of activities of daily living (ADLs), and objective physical function (see Rahman *et al.* (2004) for details on the construction of these indices). A respondent was considered in *good health* if he or she had severe disability along none or one of these dimensions. Table 10.1, which also shows the distribution of each health dimension, indicates that 68 per cent of respondents fit into the composite good health category, henceforth referred to simply as '*good health*'.

Health status models convey different information to models of the extreme event of mortality. It is likely, for instance, that mortality will depend more clearly on access to financial resources, while health reports may reflect a more broad-based set of factors. Cross-sectional models, however, introduce concerns over the direction of causality between health and some predictors. In particular, children's financial transfers are likely to be driven by parental health status rather than the reverse; for this reason, no health status models test the effects of transfers. Poor health status at the time of survey is also likely to be correlated with poor health earlier in life, which may have determined children's educational outcomes and migration opportunities. While it is unlikely that children's migration episodes in the survey year 1996 were initiated as a result of poor parental health in 1996, it is quite likely that health status in 1996 may reflect both the effects of children's migration behaviour, as well as be a cause of children's current migration patterns.

Table 10.1 Summary of health and mortality measures

Measure	Response	%	Survival	
			Yearly (%)	*Through entire follow-up (%)*
Total		100	97.8	88.6
Observed physical fitness	Yes	85	98.5	92.2
	No	15	93.7	69.0
Self-reported physical fitness	Yes	51	98.8	94.0
	No	49	96.7	83.0
Self-reported general good health	Yes	63	98.6	92.6
	No	37	96.3	81.6
Composite good health	Yes	68	98.7	93.8
	No	32	95.2	77.5
N		3,056	14,598.0	3,056

Source: MHSS (1996).

Longitudinal Models

Longitudinal models address some of the aforementioned flaws in the cross-sectional models. Logistic hazard models were constructed to predict the odds that a respondent survived a given calendar year, in terms of child characteristics, as well as controls for fixed respondent/spouse/wealth characteristics, time-varying respondent/spouse age measures, and controls for yearly mortality fluctuations.[1] Most importantly, they control for health status at the time of survey by including the three health indicator variables (self-reported general health, self-reported disability, objective disability), a strategy crucial to isolating the effects of migration on mortality net of prior interconnections between migration and health. Models of subsequent survival can also identify the effects of current transfers net of the tendency to receive more transfers when in worse health. While it is still possible that those who receive transfers are more likely to die by virtue of having been in poorer health at the time of survey, much of this relationship will be captured by the inclusion of health measures at the time of survey, which are strongly predictive of subsequent survival.

Table 10.1 demonstrates the strong relationship between health status and mortality, and thus the need to predict mortality net of health effects. After accounting for the effects of censoring from the sample due to migration, each of the three health measures is associated with at least an eleven percentage point difference in survival over the follow-up period. In particular, only 69 per cent of respondents who failed the objective physical test survived the follow-up period, compared to 92 per cent of those who passed. Smaller but substantial survival advantages were seen among those who reported physical fitness (94 per cent vs. 83 per cent) or generally good health (93 per cent vs. 82 per cent).

Each respondent contributed an observation for each person each year from 1996 to 2000 or until censoring due to death or out-migration. Results and tables display the relative increase in the odds of survival associated with a one-unit change in a predictor variable. A figure discussed at the end of the analysis converts these results into probabilities of survival for the entire follow-up period associated with the most important predictor variables when all other predictors are held at their means.

Descriptive analysis and statistical hypotheses

This section introduces predictive variables of interest, provides a descriptive overview, and introduces hypotheses derived from the theoretical literature and existing knowledge of the Bangladeshi context.

Child location and sex

It is important first to measure the effect of children's migration on respondent health. The *first hypothesis suggests that an additional international or internal migrant child, as a proportion of surviving children, should be associated with improved respondent health status and survival.* Because migration occurs in a context of patriarchal preference, and past research has shown that only migrant sons provide greater financial support, the *second hypothesis suggests that migrant sons, and not migrant daughters, will have a positive effect on respondent health and mortality.*

MHSS household rosters provide data on co-resident children, while respondents answered a separate module reporting on non-resident children. All respondents were asked to provide basic information on non-resident children, including their age, schooling, location, marital status and exchange of financial transfers. Respondents answered a question categorizing the location of each specific non-household child from among the following choices: living in the *bari* (residential compound), living in the village but not the *bari*, living in the district but not in the village, living outside the district (with a separate category identifying those living in Dhaka City), or living outside the country. For the purposes of this analysis, children living inside the district were grouped together (referred to as children living 'nearby').[2] Children living outside the district but inside Bangladesh were also grouped together and referred to as 'internal migrants'.[3] Child-specific data for co-resident and non-resident children were then tallied up to the level of the respondent, generating count variables for surviving adult (over age fifteen) children (and specifically for sons and daughters) who were either (1) in the household, (2) nearby, (3) internal migrants, or (4) international migrants. Further respondent-level count variables were constructed for the number of surviving children under age fifteen, the number of deceased children, and the number of surviving children whose location could not be identified by the respondent.

Statistical models focus on the location of a respondent's *surviving adult children* ('*children*'). All statistical models control first for the number of children, so that sex- and location-specific child counts may be interpreted as proportional to the number of children.[4] This approach separates out the effects of the intensity of children's out-migration activity from the effects of large family size on simply increasing the overall number of children who migrate. Consider, for example, respondents who each have two internal migrant sons, yet the first has eight children and the other has only two. It is important to account for the fact that the respondent with eight children has far greater opportunity to have two migrant children simply by virtue of family size. Similarly, if sex/location counts are entered without controlling for children, the effects of the first respondent's large family on health and survival will be distributed, in an unpredictable fashion, in the effects of the other sex/location-specific counts, each of which is likely to be larger for a respondent with a large family.

Table 10.2 summarizes the extent of migration among respondents' children, and differences in location by a child's sex.[5] These values again come from household rosters for co-resident children and from non-resident child

Table 10.2 Distribution of respondents' sons and daughters by location

Location	Average number of children		Having any children (%)	
	Sons	Daughters	Sons	Daughters
Out of household, in district	0.44	1.21	27	66
Internal migrant	0.60	0.52	37	34
International migrant	0.23	0.03	18	2
Adult children out of household	1.26	1.75	63	80
Adult children in household	1.17	0.44	72	34
Adult children	2.43	2.19	92	88
Children under 15	0.79	0.76	52	50
N	3,056	3,056		

Source: MHSS (1996).

modules for non-resident children. While columns 1 and 2 show that a far greater proportion of daughters (1.75 out of 2.19 total, or 80 per cent) reside away from home than sons (1.26 out of 2.43 total, or 52 per cent), most daughters remain relatively close to home (1.21 out of 2.19 total, or 55 per cent; 1.21 out of 1.75 total non-household, or 69 per cent). In contrast, a substantial proportion of total sons are internal migrants (0.60 out of 2.43 total, or 24 per cent; 0.60 out of 1.26 total non-household, or 48 per cent) or international migrants (0.23 out of 2.43 total, or 9 per cent; 0.23 out of 1.26 total non-household, or 18 per cent). Accounting for the relatively high number of surviving adult children among the study population (2.43 sons, 2.19 daughters), columns 3 and 4 show that 37 per cent of all respondents have at least one internal migrant son (15 per cent have more than one) and 18 per cent of all respondents have at least one international migrant son (5 per cent have more than one). A comparable proportion of daughters (24 per cent) and sons live outside the district, and a comparable proportion of respondents have at least one daughter living outside the district (35 per cent), but only 2 per cent of respondents have daughters living abroad.

Children's education

While very little research has addressed the impact of migrant education on the health or well-being of origin-area family members, past work does suggest a stronger effect of educational attainment on internal migrant earnings than on international migrant earnings (Massey *et al.* 1987). This is due largely to the nature of employment among the two types of migrants; internal migrants often find long-term employment with substantial returns to education, while international migrants typically find short-term employment, often at high wages, but requiring little education. *The third hypothesis anticipates a positive effect of internal migrant education on parental health/mortality, but less so for international migrants.*

Table 10.3 summarizes children's education and age, the latter of which is introduced to control for location-specific age differences between daughters, who tend to leave the household at younger ages, and sons.[6] Education level is measured in the statistical models by the schooling of the most educated child in a particular sex/location category. For instance, if a parent has three migrant sons with zero, four and eight years of schooling and no international migrant sons, then the values would be eight for internal and zero for international. The maximum is used in statistical models because it better captures the earnings potential of the child most likely to provide health-related support. If, for example, one internal migrant son completed ten years of schooling, while two others completed two years of schooling, the schooling of the best educated son (ten years) probably better captures the potential for health-related support from migrant sons than either the mean (4.7 years) or median (two years).[7] Respondents' sons have considerably more schooling than their daughters, with a 2.4-year advantage on average in terms of the most educated sibling. Migrant sons are also better educated than non-migrant sons and daughters. The average respondent's most educated internal migrant son has 7.1 years of schooling while the most educated international migrant son has 7.4 years on average (compared to 6.0 among in-household sons and 3.7 for those living elsewhere in the district). There is no statistically significant difference between the average schooling of internal and international migrants' sons.

Statistical models also account for age differences between different sex/location groups by controlling for the average age of children in each

Table 10.3 Schooling and age characteristics of respondent's children

Location	Maximum schooling		Average age	
	Male	Female	Male	Female
Out of household, in district	3.7	3.2	35.5	30.9
Internal migrant	7.1	4.1	30.7	29.7
International migrant	7.4	3.8	30.0	31.6
Total non-household	6.4	3.7	32.2	30.8
In household	6.0	5.4	26.0	20.4
Total	6.8	4.4	29.2	29.0

Source: MHSS (1996).

Note
Location-specific age/education averages shown in Table 10.3 incorporate only those respondents who had at least one child in a particular location. Sample size for all children is the same as for Table 10.2. Sample sizes for each location-specific measure equal total sample size times percent of respondents having any child in specific sex/location combination (found in Table 10.2).

group (results not shown).[8] Between-group age differences may in part drive differences in schooling or transfer activity between groups. Columns 3 and 4 show that children living outside the household tend to be older than children living in the household. Sons living inside the household tend to be substantially older on average than daughters living in-household (26.0 for sons, 20.4 for daughters) since women typically marry younger and marry outside the household. Sons living outside the household tend to be slightly older than comparable daughters (32.2 for sons, 30.8 for daughters). Not surprisingly, the average age of all living sons and daughters does not differ substantially (29.2 for sons, vs. 29.0 for daughters).

Parent–child transfer measures

A fourth, tentative hypothesis suggests that certain aspects of the relationship between children's migration and health should in fact be attributable to the extent of financial transfers received from migrant children. The hypothesis is tentative due to the difficulties inherent in estimating both the true value and meaning of transfers. The specific relationship between transfers and mortality may be lost amidst the diverse motivations for and uses of transfer income (spousal support, land investment, debt repayment), and the wide gulf between the value of transfers received in any particular year and the true strength of a parent–child transfer relationship. Take the example of two migrants. One is a circular migrant working in a factory and the other is a successful professional living in the city with his wife, yet both transfer the same amount of income in a year. The latter migrant sends money only to support his father, while the former sends money for his wife and children as well. The latter's money is invested in land accumulation, while the former's is invested in debt service and consumption. The latter earns ten times more income in any given year. Clearly the transfers made in that particular year have only limited bearing on their ability or desire to help their fathers in the long run.

Transfers received in any single year may be particularly unsuited to understand an extreme event such as mortality, which can often be prevented only through cash expenditures well beyond the means of all but the wealthiest families. It is unlikely that even the most successful migrant would transfer the $3,000 to $5,000 it might cost to treat coronary heart disease in any year but the year in which the intervention took place. In fact it is likely that successful migrants with economically self-sufficient parents would choose to make little or no transfer in typical years in order to prepare for larger transfers when they were really needed. As time passes between measurement of transfers and a longitudinal mortality observation, transfers are likely to have even less relevance. While transfers must be a primary factor mediating any relationship between children's migration and parental health, the connection could be quite difficult to establish in this context.

Table 10.4 summarizes respondent transfer activity with non-household kin. Transfer measures combine a respondent's own reported transfers with

Table 10.4 Transfer patterns

Location	Sex	Overall transfer activity		Transfers received				Transfers given			
		Any children (%)	Any transfers (%)	Any transfers received (%)	Mean transfer value ($)			Any transfers given (%)	Mean transfer value ($)		
					If any received	Overall			If any given	Overall	
Out of household, in district	Male	27	5	8	63	5		2	85	2	
	Female	66	10	8	62	5		10	22	2	
Internal migrant	Male	37	25	23	320	74		3	179	5	
	Female	34	12	8	45	4		5	138	7	
International migrant	Male	18	12	12	1,212	141		1	2,138	26	
	Female	2	1	0	275	1		1	16	0	
Total non-household	Male	63	38	36	605	221		5	655	33	
	Female	80	28	16	66	10		16	86	13	

Source: MHSS (1996).

those of his or her spouse. Compared to a measure that includes only own transfers, this measure better captures the financial support at the respondent's disposal, and tends to better predict health and survival. Analytic models actually measure transfers in terms of gross flows, calculated as the sum of transfers received and given, and net flows, or transfers received, minus transfers given.

Sons, particularly internal and international migrants, dominate transfer activity. Respondents are twice as likely to have transfers with internal migrant sons (25 per cent) than internal migrant daughters (12 per cent), and three times more likely to receive transfers (23 per cent to 8 per cent). When transfers were received, internal migrant sons gave seven times more than internal migrant daughters ($320 to $45), resulting in the average respondent receiving $74 from internal migrant sons (while giving only $5) but only $4 from internal migrant daughters (while giving $7). Even larger sex differences prevail in international migrant transfer activity. A two-thirds majority of respondents with sons living abroad had transfers (12 per cent out of 18 per cent with sons abroad), almost entirely a result of transfers received. When transfers were received from abroad, they averaged $1,212, so that the average respondent received $141 from sons living abroad, resulting in a net inflow of $115 (including $26 given on average). The average respondent received only $1 from international migrant daughters and gave a negligible amount. When all non-household children are pooled, sons were more than twice as likely to give as daughters, and the average respondent received a $178 inflow from sons ($221 received – $33 given) compared to a $3 outflow to daughters ($10 received – $13 given).

Results

The base models of health and survival shown in Table 10.5 address the first hypothesis, namely that children's migration should have a positive effect on health and survival. Controlling for the total number of children, measures of children living outside the district or outside Bangladesh tend towards positive association with good health and survival, yet their effects rarely approach a reasonable level of statistical significance. Each additional internal migrant child is associated with a 13 per cent increase in the odds of reporting good health, but the effect is significant at only the $p < 0.10$ per cent level, while a positive association with survival does not approach significance. Additional overseas children have a positive association with good health, but the effect is again insignificant. International migrant children have a positive association with survival in the absence of controls for health status at the time of survey, yet the significance of this effect does not hold up to the inclusion of the highly significant health status controls. These results are therefore inconclusive regarding potential positive effects of migration on health or survival.

The models shown in Table 10.6 test the second hypothesis that the

Table 10.5 Base models of health and mortality child counts disaggregated by location, but not by sex

Effect/standard error[a]	Odds of good health	Odds of survival – no health controls	Odds of survival – with health controls
Total living children	−0.02	0.06	0.05
	(0.06)	(0.08)	(0.07)
Total children out of household, in district	0.02	0.05	0.07
	(0.07)	(0.09)	(0.09)
Total internal migrant children	0.13**	0.14	0.12
	(0.07)	(0.09)	(0.09)
Total international migration children	0.13	0.23**	0.19
	(0.11)	(0.13)	(0.13)
Disability measured from ADL self-reports	–	–	−0.56***
			(0.19)
Bad performance on physical test	–	–	−0.81***
			(0.18)
Poor self-reported global health	–	–	−0.49***
			(0.17)
Good health index			
Constant	6.63***	8.37***	7.92***
	(1.02)	(1.35)	(1.32)
Observations	3,056	14,598	14,598
Log likelihood	309.13	410.2	521.35
DF	22	26	29

Source: MHSS (1996).

Notes
a Robust standard errors presented below coefficient.
** Significant at 5%;
*** significant at 1%.

effects of migration on health and survival should operate primarily through sons' migration. Support for this hypothesis would explain the failure to support the first hypothesis if positive effects of migrant sons were diluted by smaller effects for daughters.[9] The first column addresses cross-section models of health, finding that an additional son living outside the district is associated with a 24 per cent increase in the odds of having good health (significant at the $p < 0.01$ level), while each additional daughter living outside the district does not differ statistically from other daughters or from the reference category, namely co-resident sons. Neither international migrant sons nor daughters are associated with improved health.

Sons' migration provides even more power in predicting mortality risk in the follow-up period. After controlling for the three indicators of positive health status at the time of the survey, all of which are highly significant

Table 10.6 Models of health and mortality with child counts disaggregated by location and sex

Effect/standard error[a]	Including women's location effects			Removing women's location effects	
	Odds of good health	Odds of survival – no health control	Odds of survival – with health control	Odds of survival – no health control	Odds of survival – with health control
Total living children	−0.06 (0.07)	0.04 (0.09)	0.03 (0.09)	−0.06 (0.07)	0.02 (0.09)
Total male children out of household, in district	0.06 (0.09)	−0.05 (0.11)	−0.02 (0.11)	0.06 (0.09)	−0.00 (0.12)
Total male internal migrant children	0.24*** (0.09)	0.23** (0.11)	0.21* (0.11)	0.23*** (0.09)	0.22* (0.12)
Total male international migration children	0.13 (0.12)	0.32** (0.15)	0.30** (0.15)	0.14 (0.12)	0.30* (0.15)
Total female co-resident children	0.14 (0.12)	0.08 (0.16)	0.10 (0.15)	—	—
Total female children out of household, in district	0.04 (0.08)	0.14 (0.11)	0.17 (0.11)	—	—
Total female internal migrant children	0.06 (0.09)	0.07 (0.12)	0.06 (0.12)	—	—
Total female international migration children	0.41 (0.25)	−0.09 (0.33)	−0.26 (0.31)	—	—

	(1)	(2)	(3)	(4)	(5)
Total female children	—	—	—	0.07	0.14
				(0.07)	(0.10)
Disability measured from ADL self-reports	—	—	−0.54***	—	−0.55***
			(0.19)		(0.19)
Bad performance on physical test	—	—	−0.83***	—	−0.81***
			(0.18)		(0.18)
Poor self-reported global health	—	—	−0.50***	—	−0.49***
			(0.17)		(0.17)
Constant	6.58***	8.39***	8.04***	6.58***	7.91***
	(1.03)	(1.36)	(1.34)	(1.03)	(1.34)
Observations	3,056	14,598	14,598	3,056	14,598
Log likelihood	316.3	423.8	538.9	314.3	536.1
DF	25	29	32	23	30

Source: MHSS (1996).

Notes
a Robust standard errors presented below coefficient.
* Significant at 10%;
** significant at 5%;
*** significant at 1%.

predictors of survival, additional sons living abroad are associated with a 30 per cent increase in the odds of survival in each year, significant at the $p < 0.05$ level, while additional internal migrant sons are associated with a 20 per cent increase in annual survival odds. The internal migrant son effect is significant at the $p < 0.05$ level when no health controls are entered, and falls just below the $p < 0.05$ significance level once health controls are introduced.

The results in Table 10.6 support the second hypothesis and explain the statistical weakness of the first. While children's out-migration does indeed have positive effects on respondent health and survival, these operate exclusively through positive effects of sons' migration, suggesting that any benefits of migration on health continue to be structured by pre-existing structures of son preference. While internal migration appears to be a strong predictor of both health status and survival, international migrant sons only predict survival. The resulting parsimonious model, which includes location-specific controls for non-resident sons but not for non-resident daughters, is shown in the final two columns of Table 10.6. The magnitude of these effects is summarized in Figure 10.3 which compares the health and

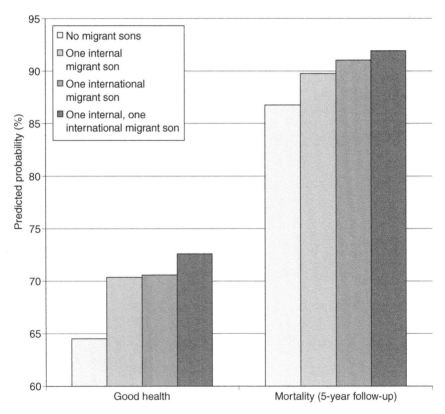

Figure 10.3 Estimated effects of migrant sons on health status and survival using Table 10.6 models (source: MHSS (1996)).

survival of respondents with no migrant sons to those with one internal migrant son, one international, or both. A comparison of health effects shows that 72 per cent of respondents with both an internal and international migrant son are predicted to be in good health, compared to 65 per cent of those with no migrant sons (again, the mean was 68 per cent). Survival effects show that even after controlling for differences in health status at the time of survey, which also favour respondents with migrant children, 90.4 per cent of respondents with one internal and one international migrant son survive the follow-up period, compared with 88 per cent of those with no migrant sons. Looking at the entire distribution of migration outcomes and ignoring the substantial association between migration and health measurements, the migration process allowed an additional two out of every 100 respondents to survive the follow-up.

The models in Table 10.7 begin to explain the pathways of the migration effect by addressing the third hypothesis, namely that migrant education can explain much of the relationship between migration and health, particularly for internal migrants. Three separate specifications incorporated the effects of the educational attainment of a respondent's most educated internal and international migrant son ('schooling of the most educated son' referred to as 'schooling'): the first included a single linear schooling term (not shown), the second included linear and square terms to account for diminishing returns to education, and the third included categorical controls for attended or completed primary school (up to five years), attended or completed secondary school (six to ten years), and attended post-secondary schooling (ten plus years). Since the schooling measure is by necessity zero if a respondent had no sons or no migrants, the third specification's reference category – no completed schooling – combines respondents who have migrant sons together with those who have migrant children who had no schooling, but differences between these two groups would still be captured by the total migrant sons measure.

Each specification finds no significant effect of migrant schooling on reported health (column 1). Models of survival find no relationship with international migrant education, but they find a strong positive association with internal migrant schooling. A linear effect of internal migrant schooling on survival is rejected in favour of either the non-linear or categorical education specification. The second specification shows a positive linear effect on survival, as well as a negative square effect. The resulting pattern reflects a peak in survival for respondents whose internal migrant sons have eight years of schooling. The third specification reflects a similar result, with only those respondents whose internal migrant sons have completed secondary education experiencing improved survival (83 per cent increased odds). Both of these education specifications reduce the association between the number of internal migrant sons and survival below the level of statistical significance, although an association with reported health status remains significant.

The first column of Table 10.8 presents the final mortality model based

Table 10.7 Models of health and mortality including continuous and categorical specifications for migrant son's education

Effect/standard error[a]	Continuous education specification		Categorical education specification	
	Odds of good health	Odds of survival	Odds of good health	Odds of survival
Total living children	-0.06	0.03	-0.06	0.02
	(0.07)	(0.09)	(0.07)	(0.09)
Total male children out of household, in district	0.06	-0.02	0.06	-0.01
	(0.09)	(0.12)	(0.09)	(0.12)
Total male internal migrant children	0.26**	0.02	0.23**	0.05
	(0.11)	(0.14)	(0.11)	(0.15)
Total male international migration children	0.16	0.10	0.19	0.12
	(0.20)	(0.28)	(0.20)	(0.29)
Total female children	0.07	0.12	0.07	0.13
	(0.08)	(0.10)	(0.08)	(0.10)
Max internal migrant male schooling	-0.05	0.18**	–	–
	(0.06)	(0.08)		
Max internal migrant male schooling – squared	0.01	-0.01*	–	–
	(0.01)	(0.01)		
Max international migrant male schooling	-0.02	-0.12	–	–
	(0.08)	(0.14)		
Max international migrant male schooling – squared	0.00	0.02	–	–
	(0.01)	(0.01)		

Max internal migrant male schooling – primary	—	—	-0.14	0.09
			(0.26)	(0.34)
Max internal migrant male schooling – secondary	—	—	-0.16	0.83***
			(0.23)	(0.31)
Max internal migrant male schooling – post-secondary	—	—	0.34	0.27
			(0.26)	(0.37)
Max international migrant male schooling – primary	—	—	-0.38	-0.15
			(0.37)	(0.48)
Max international migrant male schooling – secondary	—	—	-0.06	0.31
			(0.34)	(0.52)
Max international migrant male schooling – post-secondary	—	—	0.05	0.94
			(0.41)	(0.71)
Constant	6.73***	7.96***	6.90***	8.09***
	(1.03)	(1.29)	(1.04)	(1.32)
Observations	3,056	14,598	3,056	14,598
Log likelihood	324.1	529.3	330.0	550.3
DF	27	34	29	36

Source: MHSS (1996).

Notes

a Robust standard errors presented below coefficient.

* Significant at 10%;

** significant at 5%;

*** significant at 1%.

Table 10.8 Effects of various specifications of parent–child transfer relationships on odds of survival

Effect/standard error[a]	(1)	(2)	(3)	(4)	(5)	(6)	(7)	(8)	(9)
Total male internal migrant children	0.05 (0.15)	0.02 (0.15)	0.01 (0.15)	0.05 (0.16)	0.04 (0.15)	0.04 (0.15)	0.08 (0.15)	0.06 (0.15)	0.09 (0.15)
Total male international migration children	0.28* (0.15)	0.24 (0.15)	0.21 (0.15)	0.22 (0.18)	0.22 (0.17)	0.26* (0.15)	0.19 (0.17)	0.29* (0.15)	0.19 (0.16)
Max internal migrant male schooling – primary	0.08 (0.34)	0.03 (0.35)	0.03 (0.35)	0.07 (0.36)	0.06 (0.35)	0.07 (0.35)	0.12 (0.36)	0.10 (0.34)	0.12 (0.35)
Max internal migrant male schooling – secondary	0.86*** (0.32)	0.83*** (0.32)	0.83*** (0.32)	0.86*** (0.33)	0.85*** (0.33)	0.85*** (0.32)	0.90*** (0.33)	0.87*** (0.32)	0.90*** (0.32)
Max internal migrant male schooling – post-secondary	0.31 (0.37)	0.26 (0.37)	0.24 (0.37)	0.30 (0.37)	0.28 (0.37)	0.30 (0.37)	0.36 (0.37)	0.34 (0.37)	0.38 (0.36)
Any gross transfer activity	–	0.26 (0.17)	–	–	–	–	–	–	–
Log total gross transfers			0.03– (0.02)						
Any gross transfer activity With internal migrants			– (0.30)	0.01 (0.30)					
Any gross transfer activity With international migrants			–	0.15 (0.37)					

	(1)	(2)	(3)	(4)	(5)	(6)	(7)
Log total gross transfers	—	—	—	—	—	—	—
With internal migrants	—	—	—	—	0.01 (0.03)	—	—
Log total gross transfers	—	—	—	—	—	—	—
With international migrants	—	—	—	—	0.02 (0.04)	—	—
Any net transfer activity	—	—	—	—	—	0.06 (0.17)	—
Log total net transfers	—	—	—	—	—	—	−0.16 (0.29)
Any net transfer activity	—	—	—	—	—	—	0.21 (0.39)
With internal migrants	—	—	—	—	—	—	—
Any net transfer activity	—	—	—	—	—	—	—
With international migrants	—	—	—	—	—	—	—
Log total net transfers	—	—	—	—	—	—	—
With internal migrants	—	—	—	—	—	—	—
Log total net transfers	—	—	—	—	—	—	—
With international migrants	—	—	—	—	—	—	—
Log likelihood	538.0	530.5	532.0	540.4	544.8	537.4	535.8
DF	31	32	32	33	33	32	33

Source: MHSS (1996).

Notes
a Robust standard errors presented below coefficient.
** Significant at 5%;
*** significant at 1%.

on migration and schooling; it includes main effects for sons' location and a categorical education specification only for internal migrants. The two factors which predict mortality are international migration and whether any internal migrants have completed secondary schooling and no more. Based on the secondary schooling coefficient, a hypothetical survival projection was estimated over the entire five-year follow-up period. While the mean mortality rate of 2.5 per cent implies a 12 per cent likelihood of death over the follow-up period, respondents who have internal migrant children with secondary schooling would have only a 4 per cent likelihood of death over the follow-up period. The international migrant son coefficient would result in a 9 per cent likelihood of death through the follow-up period for respondents with one international migrant son.

The results thus far suggest that migrant children who can provide lucrative financial support may have a particularly important effect on parental survival, which is not surprising given the potentially greater role of costly medical intervention and hospital care in promoting survival among those who are ill. International migrants can provide large transfers by virtue of higher salaries in overseas labour markets; however, studies in a number of contexts suggest that educated internal migrants may actually earn more than international migrants. Yet for international migrants and educated international migrants to have a more substantial impact on parental health, these effects would most likely operate through the mechanism of increased financial support among respondents who have such migrant children. The fourth hypothesis addresses this concern, suggesting that the receipt of financial transfers can explain much of the impact of sons' migration on health and mortality.

Table 10.8 displays the results of a number of specifications relating respondent/child transfer activity to survival, in order to test the fourth, tentative, hypothesis that transfer activity would explain some of the effect of migration on improved survival. As discussed in the methodology section, cross-sectional models of health status are ignored due to concerns over reverse causation between transfers and health status in the same year. Reported transfer specifications address both net activity, to capture the direction of the transfer flow, and gross activity, to capture the strength of the overall relationship. Specifications incorporate transfer value (logged), to capture the size of the flow, and any transfer activity, to capture the presence of a flow. Specifications address both transfers from all non-resident sons, as well as separate measures specific to internal and international migrant sons.

These models, as well as a number of unreported specifications, suggest a tendency towards positive association between transfers and survival, yet none approaches statistical significance.[10] International migrant transfer measures tend to reduce the size and significance of the total international migrant sons' variable, yet none approaches significance. The models suggest that gross transfer activity, which might capture continued parental investments in children's migration activity, might serve as a better predictor of survival than net transfer receipts, yet these too fall short of standards of statistical significance. Taken together, the results fail to lend any support to the fourth hypothesis, suggesting two possible explanations. First, migrant sons may genuinely enhance respondent survival through non-financial pathways such as improved health care knowledge, or through a spurious association with parental traits that may both facilitate children's migration and predict continued survival. A more likely conclusion suggests that transfer measures taken at one point in time inadequately represent the true mechanisms linking transfers to survival, such as a migrant son's capacity to raise vast sums of money to pay for particular medical interventions. The volume of migrant sons' transfers, even considering their role in supporting spouses, children and long-term investments, suggest that they must have some effect on respondent health.

Conclusion

The preceding results add weight to the argument that migration, whether internal or international, may benefit not only the wealth of left-behind populations, but also their health and survival. These results sit well with an emerging model of migration in which elders in LDCs facilitate children's migration activities, partly as a means of paying for their own retirement, and control many of the resources required by returning migrants. We must therefore situate these results in the broader context of migration within Asia, where both internal and overseas destinations offer limited migrant settlement opportunities. The international migration effects hold a particular logic in light of numerous studies, some in this volume, which contrast immigrant health deficits in Asia against health surpluses among migrants to societies where the right to health care and/or settlement is guaranteed. If migration under these circumstances generates financial gains, most of which are remitted to the country of origin, these gains might logically lead to both diminished health for movers and improved health for stayers.

Yet if financial transfers are to explain the relationship between movers' sacrifices and stayers' benefits, then the most important contribution of this research may be the non-finding that no measure of parent–child transfer activity statistically captures migration's impact on respondent survival. The second, well-supported hypothesis that only migrant sons would affect respondent health was in fact spurred by research demonstrating the predominance of sons' transfers over daughters', yet these substantial transfers could not in fact explain the migration/health relationship. This raises the crucial methodological point that it is quite difficult to construct a meaningful model relating transfers to health, given the multidirectional pathways linking these processes. A model relating migrant transfers in one year to parental health outcomes in another would require two survey rounds and substantial econometric skill. Yet the straightforward models presented here often leave more questions unanswered than answered.

Fortunately, the genuine concern of migration/health research is not about transfers in particular, but about the complex social, economic and demographic processes resulting in transfers. The current work raises important questions in this regard. For instance, is the relationship between the secondary education of internal migrants and parental health driven by migrants' need to secure their own futures in the rural area? If so, would further disaggregation reveal groups of vulnerable respondents – possibly women, the very elderly or the landless – that are negatively impacted by migration? Two concerns also unfold from the fact that these results may be driven in part by the hierarchical, dependent relationship between parents and children. First, would these results differ if we studied the health of migrants' wives rather than parents? Second, what long-term, non-monetary effects stem from the migration process, or from processes of transnational-ization that encompass the migration process itself? Chapter 11 (this volume) demonstrates both of these concerns in finding that the wives of migrants from rural India to Delhi experience diminished health due to increased HIV exposure, yet migration could also yield long-term effects on the elderly through pathways such as dietary change.

Finally, these results find that migration is of net benefit to the parents of migrants. Limitations to government programmes or markets ensuring livelihood diversification, long-term investment or health finance in some of Asia's poorest nations makes children's migration, even under harsh circum-stances, an opportunity for enhanced health, wealth and well-being for those elders who can afford to invest in children's education and overseas travel. Perhaps a more fruitful angle on the migration/health question would be to ask: 'What about the health of those who do not have, or cannot afford, migrant children?'

Notes

1 Models also incorporate a Huber-White correction for intra-cluster correlation in the distribution of the probability of mortality across observation years for a given respondent.

2 Early models distinguished children living in the same village or residential compound from children living in more distant parts of the same district, and found no statistical difference between the two groups in terms of predicting respondent health or mortality, so these distinctions were not employed.

3 A majority of male migrants to other districts move to smaller cities (such as Chittagong, Narayanganj, and Sylhet), and should thus be considered as a similar group to migrants to Dhaka. Previous models (not shown) treated internal migrants to Dhaka City and other internal migrants as separate groups, and found no statistical difference between the groups in terms of their effect on health or survival.

4 All models also include controls (not presented or shown in the tables) for sur-viving children under age fifteen, deceased children, and children whose location cannot be identified by the respondent.

5 Given the small percentage of cases censored due to mortality and migration, child location distributions for each person-year observation in longitudinal

models remain similar to within 1 per cent of the comparable respondent-level values.

6 Given rapid increases in schooling over time, particularly for girls, age controls account for the possibility that the younger group of non-resident daughters may also have higher levels of educational attainment.

7 In reality, such disparities among sons rarely exist and the modal numbers of children in each sex/location combination are 0 and 1, rendering the maximum/median/mean question moot. Maximum, mean and median schooling are highly correlated for each sex/location combination. Preliminary models (not shown) confirmed that while median and mean education bore a statistically significant relationship to parental health and survival, the education maximum offered improved statistical power.

8 For respondents who have no children in a particular sex/location category, average age is set to the mean for the whole population in Table 10.3 and in the statistical model 1.

9 Separate models also addressed distinctions between married and unmarried sons and daughters living abroad and elsewhere in Bangladesh (not shown). These models found no statistically significant differences between married and unmarried children within each location/sex category.

10 Unreported specifications include transfers given and received; alternate transformations of transfer value (including untransformed and square-root), and separate measures of positive and negative net transfer flow.

References

Afsar, R. (1999) 'Rural–urban dichotomy and convergence: emerging realities in Bangladesh', *Environment and Urbanization*, 11, 1 (April): 235–46.

Arthur, W.B. and McNicoll, G. (1978) 'An analytical survey of population and development in Bangladesh', *Population and Development Review*, 4, 1 (March): 23–80.

Asis, M.M.B., Domingo, L., Knodel, J. and Mehta, K. (1995) 'Living arrangements in four Asian countries: a comparative perspective', *Journal of Cross-Cultural Gerontology*, 10: 145–157.

Becker, G.S. and Tomes, N. (1976) 'Child endowments and the quantity and quality of children', *Journal of Political Economy*, 84: S143–62.

Boserup, E. (1965) *The Conditions of Agricultural Growth*, Chicago, IL: Aldine Press.

Cain, M., Syeda, R., Khanam, S.N. and Shamsun N. (1979) 'Class, patriarchy, and women's work in Bangladesh', *Population and Development Review*, 5, 3 (September): 405–38.

Coleman, J.S. (1988) 'Social capital in the creation of human capital', *American Journal of Sociology*, 94: 595–620.

Coleman, J.S. (1993) 'The Rational Reconstruction of Society', *American Sociological Review*, 58, 1: 1–15.

Cowgill, D.O. (1986) *Aging Around the World*, Belmont, CA: Wadsworth.

Durand, J., Kandel, W., Parrado, E.A. and Massey, D.S. (1996) 'International migration and development in Mexican communities', *Demography*, 33, 2: 249–64.

Faist, T. (2000) *The Volume and Dynamics of International Migration and Transnational Social Spaces*, New York: Clarendon Press.

Faveau, V. (1994) 'Matlab: physical setting and cultural background', in V. Faveau (ed.) *Matlab: Women, Children, and Health*, ICDDR, B Special Publication No. 35, Dhaka, Bangladesh: Pioneer, pp. 13–28.

Frank, R. and Hummer, R.A. (2002) 'The other side of the paradox: the risk of low birth weight among infants of migrant and nonmigrant households within Mexico', *International Migration Review*, 36, 3: 746–65.

Frankenberg, E. and Kuhn, R. (2004) 'The role of social context in shaping intergenerational relations in Indonesia and Bangladesh', in M. Silverstein, R. Giarusso and V. Bengston (eds) *Intergenerational Relations Across Time and Place*, New York: Springer.

Frankenberg, E. and Thomas, D. (2000) 'The Indonesia family life survey (IFLS): study design and results from Waves 1 and 2', DRU-2238/1. NIA/NICHD. RAND.

Harris, J.R. and Todaro, M.P. (1970) 'Migration, unemployment and development: a two sector analysis', *American Economic Review*, 60, 1: 126–42.

Hermalin, A.I. and Myers, L. (2002) 'Aging in Asia: facing the crossroads', in A.I. Hermalin (ed.) *The Well-Being of the of the Elderly in Asia*, Ann Arbor: University of Michigan Press, pp. 1–24.

Hummer, R.A., Rogers, R.G. and Eberstein, I.W. (1998) 'Sociodemographic differentials in adult mortality in the United States: review of analytic approaches', *Population and Development Review*, 24, 3: 553–78.

Iwasaki, M., Otani T., Sunaga, R., Miyazaki, H., Xiao, L., Wang, N., Yosiaki, S. and Suzuki, S. (2002) 'Social networks and mortality based on the Komo-Ise cohort study in Japan', *International Journal of Epidemiology*, 31, 6: 1208–18.

Kabir, Z.N., Szebehely, M. and Tishelman, C. (2002) 'Support in old age in the changing society of Bangladesh', *Ageing and Society*, 22: 615–36.

Kanaiaupuni, S.M. and Donato, K.M. (1999) 'Migradollars and mortality: the effects of migration on infant survival in Mexico', *Demography*, 36, 3: 339–53.

Knodel, J., Chayovan, N. and Siriboon, S. (1992) 'The impact of fertility decline on familial support for the elderly', *Population and Development Review*, 18: 79–103.

Kuhn, R. (1999) 'The logic of letting go: individual and family migration from rural Bangladesh'. Unpublished doctoral dissertation, University of Pennsylvania.

Kuhn, R. (2001) 'Never far from home: parental assets and migrant transfers in Matlab, Bangladesh', RAND Labor and Population Working Papers, 2001–12.

Kuhn, R. (2003) 'Identities in motion: social exchange networks and rural–urban migration in Bangladesh', *Contributions to Indian Sociology*, 37, 1–2: 311–37.

Kuhn, R. and Menken, J. (2002) 'Migrant social capital and education in migrant-sending areas of Bangladesh: complements or substitutes?', Population Association of America, Atlanta, GA.

Lee, Y.J., Parish, P. and Willis, R. (1994) 'Sons, daughters, and inter-generational support in Taiwan', *American Journal of Sociology*, 99: 1010–41.

Lillard, L.A. and Willis, R.J. (1994) 'Intergenerational educational mobility: effects of family and state in Malaysia', *Journal of Human Resources*, 29: 1126–67.

Litwak, E. (1960) 'Geographic mobility and extended family cohesion', *American Sociological Review*, 25: 385–94.

Lucas, R.E.B. and Stark, O. (1985) 'Motivations to remit: evidence from Botswana', *Journal of Political Economy*, 9: 1–18.

Mason, K.O. (1992) 'Family change and support of the elderly in Asia: what do we know?', *Asia-Pacific Population Journal*, 7: 13–32.

Massey, D.S. and Espinosa, K. (1997) 'What's driving Mexico–US migration: a theoretical, empirical and policy analysis', American *Journal of Sociology*, 102: 939–99.

Massey, D.S. and Parrado, E.A. (1994) 'Migradollars: the remittances and savings of Mexican migrants to the United States', *Population Research and Policy Review*, 13: 3–30.

Massey, D.S., Alarcon, R., Durand, J. and Gonzales, H. (1987) *Return to Axtlan: The Social Process of International Migration from Western Mexico*, Berkeley: University of California Press.

Massey, D.S., Arango, J., Hugo, G., Kouaouci, A., Pellegrino, A. and Taylor J.E. (1999) *Worlds in Motion: Understanding International Migration at the End of the Millennium*, New York: Oxford University Press.

Menken, J. and Phillips, J.F. (1990) 'Population change in a rural area of Bangladesh, 1967–87', *Annals of the American Academy of Political and Social Science*, 510: 87–101.

Mostafa, G., Kashem Shaikh, M.A., van Ginneken, J.K. and Sarder, A.M. (1998) 'Demographic surveillance system – Matlab, registration of demographic events, 1996', Volume 28, Dhaka: ICDDR, B.

National Institute of Population Research and Training (NIPORT), Mitra and Associates (MA) and ORC Macro (ORCM) (2001) *Bangladesh Democratic and Health Survey 1999–2000 Dhaka, Bangladesh and Calverton*, Maryland: NIPORT, MA and ORCM.

Parsons, T. (1943) 'The kinship system of the contemporary United States', in T. Parsons (1954) *Essays in Sociological Theory*, Glencoe: The Free Press, pp. 177–96.

Portes, A. (1996) 'Transnational communities: their emergence and significance in the contemporary world-system', in R.P. Korzeniewicz and W.V.C. Smith (eds) *Latin America in the World-economy*, London: Greenwood Press.

Portes, A. and Sensenbrenner, J. (1993). 'Embeddedness and immigration: notes on the social determinants of economic action', *The American Journal of Sociology*, 98, 6: 1320–50.

Rahman, M.O. (1999) 'Family matters: the impact of kin on the mortality of the elderly in rural Bangladesh', *Population Studies*, 53: 211–25.

Rahman, M.O. and Barsky, A.J. (2003) 'Self-reported health among older Bangladeshis: how good a health indicator is it?', *The Gerontologist*, 43: 856–63.

Rahman, M.O., Foster, A. and Menken, J. (1992) 'Older widow mortality in rural Bangladesh', *Social Science and Medicine*, 34, 1: 89–96.

Rahman, O., Menken, J. and Kuhn, R. (2004) 'The impact of family members on self-reported health of elderly men and women in rural Bangladesh', *Ageing and Society*, forthcoming.

Rahman, O., Menken, J., Foster, A., Peterson, C., Khan, M., Kuhn, R. and Gertler, P. (1999) *The 1996 Matlab Health and Socioeconomic Survey: Overview and Users' Guide*, DRU-2018/1. RAND.

Rogers, A. and Castro, L.J. (1981) 'Age patterns of migration: cause-specific profiles', *IIASA Reports*, 4, 1: 125–59.

Shah, N.M. and Menon, I. (1999) 'Chain migration through the social network: experience of labour migrants in Kuwait', *International Migration*, 37, 2: 361–82.

Simmons, R. (1996) 'Women's lives in transition: a qualitative analysis of the fertility decline in Bangladesh', *Studies in Family Planning*, 27, 5 (September–October): 251–68.

Skinner, G.W. (1997) 'Family systems and demographic processes', in D.I. Kertzer and T. Fricke (eds) *Anthropological Demography: Toward a New Synthesis*, Chicago: University of Chicago Press, pp. 53–95.

Stark, O. (1982) Research on rural-to-urban migration in LDCs: the confusion frontier and why we should pause to rethink afresh', *World Development*, 10, 1: 63–70.

Stark, O. and Bloom, D.E. (1985) 'The new economics of labour migration', *American Economic Review*, 75: 173–8.

United Nations Population Division (2002) *International Migration Report*, ESA/P/WP,178, November.

United Nations (2003) *World Urbanization Prospects: The 2003 Revision,* United Nations, Department of Economic and Social Affairs, Population Division.

Wu, Z.H. and Rudkin, L. (2000) 'Social contact, socioeconomic status, and the health status of older Malaysians', *Gerontologist*, 40, 2: 228–34.

11 Reproductive health status of wives left behind by male out-migrants

A study of rural Bihar, India

Archana K. Roy and Parveen Nangia

Introduction

Reproductive health is not only concerned with safe motherhood, wanted pregnancies, reproductive tract infections (RTIs) and sexually transmitted infections (STIs), but also with health matters related to contraception. The World Health Organization defines reproductive health as 'a complete physical, mental and social well-being and not merely the absence of disease or infirmity, in all matters related to the reproductive system and its functions and process' (United Nations 1995: 30). The International Conference on Population and Development held in Cairo in 1994 stressed the implications of women's health, especially their reproductive health, for overall economic development. Thus, there is a growing concern about reproductive morbidity among poor women in developing countries (United Nations 1995: 30).

In recent years migration has become a new focus of reproductive health care programmes in developing countries. The main reason for this concern is that AIDS and other sexually transmitted diseases (STDs) spread faster in a population when it becomes more mobile (Armstrong 1995; Decosas *et al.* 1995; Population Report 1998). In the special session of the General Assembly of the United Nations on HIV/AIDS held in June 2001, the declaration of commitment was: 'By 2005, develop and begin to implement national, regional and international strategies that facilitate access to HIV/AIDS prevention program for migrants and mobile workers, including the provision of information on health and social services' (cited in UNAIDS 2002: 114).

The International Organization for Migration (IOM) and the United Nations programme on AIDS (UNAIDS) recognize that AIDS and migration are two salient features of the latter half of the century. Previously, the main concern of the governments of migrant-receiving countries was that incoming migrants may bring HIV with them to their areas of destination, and while this may still be true (UNAIDS and IOM 1998: 446), it would be incorrect to assume that migrants generally only bring AIDS with them, or that they are already infected with HIV at the time of their arrival.

Few studies have been conducted on the prevalence of STDs among migrant groups (De Schryver and Metheus 1991), but research shows that the spread of HIV often coincides with migration patterns. According to United Nations' estimates, approximately 16 million people migrate each year from rural to urban areas in developing countries, excluding China (Population Report 1998). For many people in rural areas, the alternative is to seek a living outside the village. The push factors from the villages are more forceful than the promise of a good life in the cities (Todaro 1976). Such migration is generally restricted to men; women are left behind. This pattern of male migration without families is particularly prevalent in African, Latin American and Asian Countries (Population Report 1998).

Separated from their families and regular sex partners for long periods, male migrants may encounter loneliness or even a sense of isolation in a country or region where the language and cultural practices are alien (Haour-Knipe (2000), cited in Population Reference Bureau 2001: 19). Carlier (1999) found that migrants who are settled with their families often had less risky sexual behaviour than occasional travellers. Some migrants often become part of a new peer group, including a new sexual network. In a recent study, Cruz and Azarcon (2000) showed that many migrant workers are under thirty years of age and sexually active. Many are single or have left their families back home. Young, alone and socially isolated, they tend to seek comfort in intimate relationships developed while away from their families, or engage in casual or paid sex. The incidence of condom use among migrant workers is low due to poor accessibility, hesitation in buying condoms, uncertainty about the protection they provide, and reluctance to use them in intimate or steady relationships (Cruz and Azarcon 2000). Thus the disruption of social ties and family life that occur during moves, especially in situations of poverty and crisis, also increases the risk of STIs as migrants find new sex partners. Migrants become carriers of STIs and make their wives vulnerable when they visit their homes (in their own villages). This suggests that the prevalence of RTIs/STIs may be higher among the left-behind wives of migrant men than among the wives of non-migrant men. Figure 11.1 illustrates the conceptualization of the problem.[1]

The presence of an STI increases the likelihood of acquiring HIV infection. Some STIs increase the replication of HIV (Inversen et al. 1998; Population Report 2002; Rasmussen et al. 1997; Theus et al. 1998). People with genital ulcers are at two to five times greater risk of acquiring HIV/AIDS. Diseases such as chlamydia, gonorrhoea and trichomoniasis that do not cause ulcers increase the risk of HIV transmission to women by three to five times (Population Report 1993). Genital inflammation may cause microscopic cuts that can allow HIV to enter the body. Diseases causing vaginal and urethral inflammation are far more common than genital ulcers and may be responsible for a larger share of HIV transmission. Empirical studies present evidence of HIV transmission from urban to rural areas. Over time, the epidemic moves into rural populations, although rural rates remain below

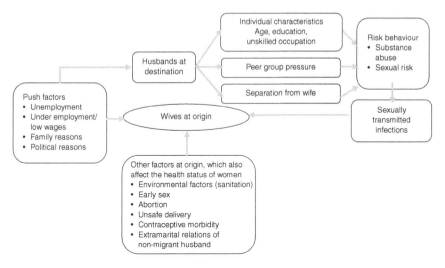

Figure 11.1 Conceptual framework.

urban rates, except in the highest prevalence countries (Gilbert 1997; UNAIDS 2002).

The Indian context

In India, as in other developing countries, migration is a result of wide regional disparity. Most of the economic activities are concentrated in a few pockets and urban centres. The fifty-fifth round of the National Sample Survey (NSS) conducted in 1999 to 2000 (Government of India 2001) showed that the male unemployment rate was higher in urban areas (9.4 per cent) compared to rural areas (7.2 per cent), but the male underemployment rate[2] was higher in rural areas (10.5 per cent) than in urban areas (6.1 per cent). There was also a remarkable difference between rural and urban wage rates. For males, the daily wage rate was Rs.44.84 in rural areas, and Rs.62.26 in urban areas (Government of India 2001). Thus, due to higher underemployment in rural areas and higher urban wages, men are pushed from rural areas, particularly areas located in less developed states, to the urban centres of more developed states.

At the end of 2001, the adult HIV prevalence rate in India was under 1 per cent. It was estimated that 3.97 million people were living with HIV/AIDS – more than in any other country besides South Africa (UNAIDS 2002: 30). The epidemic was spreading in the general population and beyond the group of high-risk behaviour. The HIV/AIDS map prepared by the National AIDS Control Organization on the basis of surveillance data showed that the prevalence of these diseases was more widespread in metropolitan areas and the industrially developed states of Maharashtra, Gujarat and Andhra Pradesh, which attract workers from all over India, particularly from states with lower income levels, such as Bihar, Uttar Pradesh, Madhya Pradesh and Orissa. Some of these low-income states, in which out-migration exceeds in-migration, have lower rates of HIV infection than the more developed states, where in-migration exceeds out-migration (UNAIDS 2002).

A recent unpublished study by Gupta and Singh (2002) on the behaviour of single male migrants in Surat, an industrial centre in India, found that migrants living in slums were more prone to indulge in risky behaviour than local residents. Similarly, Mishra (2002) in his study of 'with family' and 'without family' migrants in a Delhi slum, revealed that more than half of the migrants living without family approve of having extramarital relations where the temporary absence of the wife may be a contributing factor. Other than separations from regular sex partners, slum environments, peer group influence, the availability of disposable income and sex avenues, including commercial sex workers in slum areas, emerged as important factors motivating people to take a higher risk.

Two recent large-scale sample surveys conducted in India show that rural women have very poor knowledge of HIV/AIDS (Table 11.1). According to the second National Family Health Survey (NFHS-2) conducted in 1998 to 1999, 40 per cent of women in India and 12 per cent of women in Bihar have heard about AIDS (International Institute for Population Sciences and ORC Macro 2001a, 2001b) The Reproductive and Child Health Survey, also conducted in 1998 to 1999, put these figures at 42 and 15 per cent respectively. Both surveys found a higher prevalence rate of RTI/STI among ever-married women in Bihar, compared to the country as a whole. The NFHS-2 found that 39 per cent of women in India, and 44 per cent of women in Bihar, suffered from RTI/STI (International Institute for Population Sciences and ORC Macro 2001a, 2001b), whereas according to the RCH, the respective rates were 30 and 38 per cent (Government of India 1999).

It is also important to examine health programmes and policies in India in order to get a more comprehensive idea about the country's health system. The modern health care system in India is a three-tier system, consisting of community-based primary health care, hospital-based secondary health care, and hospital-based tertiary health care. The National Health Policy of 1983 (NHP 1983) was a major breakthrough in health policy and planning, which emphasized a preventive, promoting and rehabilitative health care approach (Nanda 2002). It recommended a decentralized system of health care through the expansion of the private curative sector. Although the NHP 1983 recommended a comprehensive health care system, under the influence of the ICPD–Cairo agenda (Nanda 2002), India committed itself to implementing the reproductive and child health approach.

Aggregate expenditure in the health sector in India is currently 5.2 per cent of GDP (Tidco-India Policies 2003), and the share of public health

Table 11.1 A comparative picture of knowledge and prevalence of reproductive morbidities among currently married women in India and Bihar by NFHS and RCH surveys

	NFHS-2		RCH	
	India (%)	*Bihar (%)*	*India (%)*	*Bihar (%)*
Knowledge about HIV/AIDS	40.3	11.7	41.9	15.1
Knowledge about RTI	–	–	45.4	67.0
Knowledge about STI	–	–	28.8	38.9
With any reproductive health problem	39.2	44.2	29.7	37.7
Percent with any abnormal white discharge	30.2	33.7	–	–
Vaginal discharge with itching or irritation	17.3	19.3	–	–
Vaginal discharge with bad odour	11.5	15.8	–	–
Vaginal discharge with severe lower abdominal pain[a]	18.7	22.8	–	–
Vaginal discharge with fever	8.1	10.1	–	–
With symptoms of a urinary track infection[b]	17.8	25.7	–	–
With abnormal vaginal discharge or symptom of UTI[b]	35.9	42.2	–	–
With painful intercourse (often)	12.5	11.4	–	–
With bleeding after intercourse (ever)[a]	2.3	2.4	–	–
Total sample of currently married women	83,649	6,661		

Source: NFHS-2 (1998–99) and RCH (1998–99).

Notes
a Not related to menstruation.
b Includes pain or burning while urinating, or more frequent or difficult urination.

expenditure is only 17 per cent of aggregate expenditure. The current annual per capita public health expenditure is no more than Rs.200 (Tidco-India Policies 2003). Hence, the reach and quality of public health services has been below a desirable standard. There is a wide disparity in public health facilities and health standards across the states and rural–urban areas.

Bihar is one of the poorer performing states in terms of various health indicators. Although Bihar spends nearly 5 per cent of its total budget on medical and public health, it has not been in a position to meet the basic health needs of its population. There are 328 hospitals in the state, of which ninety are in the private and voluntary sectors. Most of the hospitals, on account of their referral nature, are located in the urban areas. On average 260,000 people depend on one hospital, and there is one bed for every 3,000 people. Thus, the public health infrastructure is grossly inadequate.

According to the NFHS-2 (1998 to 1999) report (International Institute for Population Sciences and ORC Macro 2001a), in rural areas more than 90 per cent of households use private medical services, and only 1 per cent use community health centres/rural hospitals or primary health centres. In 2001 a new national health policy – NHP-2001 – was proposed. Its principal objective is to reduce inter-regional and urban–rural inequities and to give disadvantaged sections of society fairer access to public health services. Within these broad objectives, NHP-2001 endeavours to achieve time-bound goals, including an integrated system of surveillance, national health accounts and health statistics by 2005, zero-level growth of HIV/AIDS by 2007, and an increase in health sector expenditure by government to 6 per cent of GDP by 2010.

Objective and hypothesis

The out-migration of single males from Bihar and the existing poor health care system in the state is the focus of this chapter. The aim is to explore the reproductive health status of the left-behind wives of migrant workers in rural areas. Bihar is located in the eastern part of India. More specifically, the study examines the general morbidity pattern, reproductive health status and knowledge of RTIs/STIs, with special reference to HIV/AIDS among the left-behind wives of male out-migrants of Bihar. For comparison pur-poses, a group of women whose husbands had not migrated was also studied. It was hypothesized that reproductive morbidity would be higher among the left-behind wives than among the wives of non-migrants.

Methodology

The study was conducted in two selected districts, Rohtas and Siwan, which have high out-migration rates, but differ in levels of development (Figure 11.2). The highest sex-ratio blocks from both districts were selected. All the villages in these blocks, excluding very small villages (less than 150 house-holds), were plotted on a geographical map, and the cluster with the maximum number of villages was selected from each block. Four villages from each cluster were selected for survey, based on the information pro-vided by key informants about the major pockets of out-migrant house-holds. In one block, one more village was later added to complete the target sample size.

A complete house-listing was undertaken in the selected villages, and households were identified as migrant (where at least one male member had out-migrated for employment purposes) and non-migrant (where no male member had out-migrated for employment). A systematic random sampling technique was used to select households. In the migrant households, the cri-teria adopted for participation in the survey were duration of migration (at least one year), marital status and age of woman (currently married and aged fifteen to forty-five), and duration of marriage (at least one year). If the selected household did not match these criteria it was replaced by another household. If there was more than one woman in the household who fitted

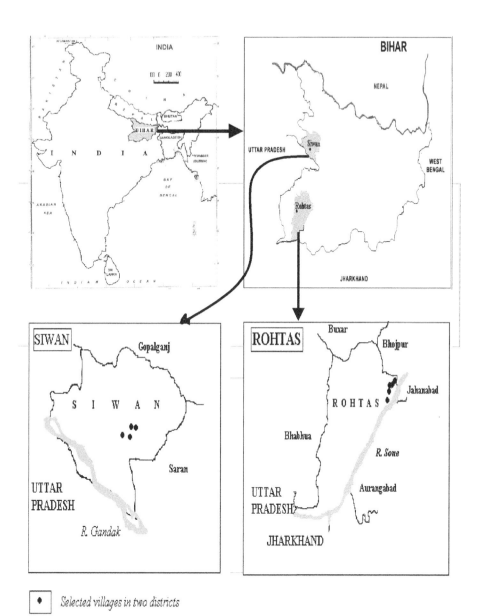

Figure 11.2 Location map of study area (source: Based on Census of India, 2001 (downloaded from www.censusindia.net)).

the selection criteria, only one was selected for the study. Thus, the left-behind wives were those whose husbands had been migrants for at least a year. For reference group purposes, a group of wives of non-migrants was selected with similar socio-economic conditions to assess the impact of male out-migration on the health of left-behind wives. The criteria for selecting these women were similar, that is, aged fifteen to forty-five and at least one year of marital duration. In all, 354 left-behind wives of migrants and 164 wives of non-migrants were interviewed in 2001 from the nine migration-prone villages of two districts in the state of Bihar, one of the major areas of male out-migration in India.

Data were collected on general and reproductive morbidity and symptoms (Table 11.2). The list of symptoms was generated on the basis of information obtained from existing literature and verified with the help of physicians. All the women were asked to report if they had any of the symptoms. The study was based entirely on self-reported symptoms and no clinical examination was conducted for the verification of the diseases. This was a major limitation of the study because self-reported symptoms need to be clinically verified. Moreover, sometimes an RTI or an STD may not have symptoms, and therefore these diseases can be under-reported.

The reference period for these morbidities was taken as six months prior to the survey. The questions were asked in the local language, *Bhojpuri*, using local terminology. Some additional information was collected from the women related to their demographic characteristics (age, education, children ever born, number of spontaneous and induced abortions), use of contraception (including method), duration of marriage, and socio-economic characteristics (caste, size of landholding, household possessions and amenities, and employment status). The women were also asked about the lifestyle factors of their husbands that could lead to risk-taking behaviour or neglect of their wives (e.g. alcohol consumption, smoking, having casual sexual relations or extramarital relations, and other habits such as chewing

Table 11.2 Symptoms of general health problems and RTIs/STIs (asked in the study)

• Headache	• Burning sensation or pain or difficulty in passing urine
• Backache	• Itching or irritation around the vagina
• Burning sensation in palms/feet and head	• Sores on genitals
• Itching in the body	• Pain during intercourse
• Dizziness	• Some protruding mass coming out from vagina
• Weakness	• Abnormal vaginal discharge (white discharge)
• Other problems	• Any menstrual problem (severe abdominal pain during periods, irregular periods, i.e. shorter or longer periodic cycle, and heavy flow or scanty flow)

Source: Fieldwork, 2001.

paan masala and tobacco, and gambling). Information on contraceptive use, particularly on the use of sterilization and intrauterine devices, which may also cause RTIs, was obtained to provide more insights into reproductive health. In developing countries mortality and morbidity due to reproductive tract infections and sexually transmitted infections is very high relative to other health problems. Because women can be reluctant to report their problems due to inhibition or social ostracism of reproductive morbidity, every effort was made to obtain accurate information from them.

The effect of various socio-economic and behavioural determinants on reproductive morbidity was analysed using a logistic regression analysis. To know how different variables affect the reproductive morbidity of women in different groups, separate analysis was undertaken for the left-behind wives of migrant men, wives of non-migrant men, and all the women together (non-migrant as well as left-behind wives). The dependent variable selected in the study was 'any reproductive morbidity', and the predictor variables were standard of living, marital duration, age of the respondent, use of contraception, respondent's knowledge/perception of her husband's extramarital relations, consumption of alcohol and chewing tobacco or *paan masala* (considered as stimulating factors), and habits such as gambling that may lead to family neglect and diversion of financial resources.

Findings and discussion

Of 2,724 households in the nine villages under study, 1,597 households had at least one male out-migrant for a minimum period of one year. Further, the migrants from 91 per cent of such households were interstate migrants, with destinations mainly in Punjab, Haryana, Delhi, West Bengal (particularly Calcutta), Gujarat and Maharashtra (particularly, Mumbai) (Figure 11.3). Among the interstate migrants, 51 per cent had migrated without their families, and the rest were either never married (17 per cent) or had migrated with their families (32 per cent). Thus the majority of the interstate male out-migrants migrated without family.

Characteristics of women and the life-style of husbands

More than 80 per cent of the women surveyed belonged to landless or marginal households[3] (Table 11.3). The left-behind wives were younger and had shorter marriages than the wives of non-migrant men. The mean age of left-behind wives was 27.9 years, compared to 29.5 years for wives of non-migrant men. A significantly higher proportion of the left-behind wives (40 per cent) had married during the past ten years, compared to 27 per cent of the wives of non-migrating men. The mean age at marriage of the women was very low (under fourteen years) and there was hardly any difference in the age at marriage between the two groups of women. The mean number of children ever born to left-behind wives (3.1) was one child less than the

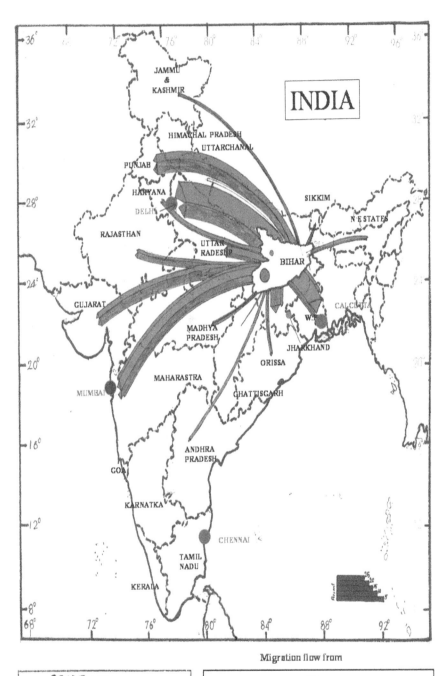

INDIA

JAMMU & KASHMIR

HIMACHAL PRADESH
UTTARCHANAL

PUNJAB

HARYANA

DELHI

SIKKIM

N E STATES

RAJASTHAN

UTTAR RADESH

BIHAR

GUJARAT

CALCUTTA

W.B

MADHYA PRADESH

JHARKHAND

ORISSA

MUMBAI

MAHARASTRA

CHATTISGARH

ANDHRA PRADESH

GOA

KARNATKA

CHENNAI

TAMIL NADU

KERALA

Migration flow from

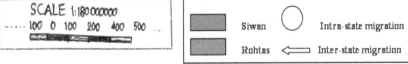

SCALE 1:180 000000
100 0 100 200 400 500

Siwan   ◯ Intra-state migration

Rohtas   ⇐ Inter-state migration

Table 11.3 Percentage distribution of women by their background characteristics

Background characteristics	Wives of non-migrant men (%)	Wives left behind by migrant men (%)	All women (%)
Caste/community			
Upper caste (Hindu)	10.4	13.0	12.2
Backward caste (Hindu)	53.0	46.0	48.3
Scheduled caste (Hindu)	21.3	17.5	18.7
Muslims	15.2	23.4	20.8
Size of landholding			
Less than one acre	83.5	81.1	81.9
More than one acre	16.5	18.9	18.1
Standard of Living Index (SLI)			
Low	34.1	29.9	31.3
Medium	42.7	39.5	40.5
High	23.2	30.5	28.2
Type of family***			
Staying with in-laws (Joint)	47.0	75.7	66.6
Staying without in-laws (Nuclear)	53.0	24.3	33.4
Age of respondent*			
15–19	4.9	11.6	9.5
20–24	19.5	26.3	24.1
25–29	24.4	20.3	21.6
30–34	27.4	23.7	24.9
35–39	15.2	10.7	12.2
40–44	8.5	7.3	7.7
Mean age (years)**	29.5	27.9	28.4
Age at marriage (years)	13.4	13.6	13.6
Age at starting cohabitation (year)	14.2	14.2	14.2
Literacy*			
Illiterate	81.1	74.3	76.4
Literate	18.9	25.7	23.6

continued

Figure 11.3 Destination of total male out-migrants from selected villages of Siwan and Rohtas (source: Fieldwork, 2001).

Note
Destinations with less than five migrants are not shown.

Table 11.3 continued

Background characteristics	Wives of non-migrant men (%)	Wives left behind by migrant men (%)	All women (%)
*Marital duration***			
Less than ten years	27.4	40.1	36.1
Eleven to twenty years	48.8	40.4	43.1
More than twenty-one years	23.8	19.5	20.8
*Children ever born (CEB)****			
≤3 CEB	39.0	61.9	54.6
≥4 CEB	61.0	38.1	45.4
Mean CEB	4.2	3.1	3.2
*Experienced abortion***			
No	78.7	85.6	83.4
Yes	21.3	14.4	16.6
Work in own or others' field	55.5	39.5	44.6
Work for wages**	42.7	29.4	33.6
Ever used contraception***	28.0	20.3	22.8
Ever use of contraception@ (sterilization or IUD)***	20.7	9.6	13.1
Duration of migration			
Up to five years	–	34.5	–
Six to ten years	–	30.5	–
More than ten years	–	35.0	–
Habits of husband			
Drinking habit of husband**	42.1	32.5	35.5
Smoking habit of husband	26.8	29.4	28.6
*Extramarital relations of husband****			
Yes	7.9	12.7	11.2
Cannot say	7.3	20.6	16.4
No	84.8	66.7	72.4
Any other habit (gambling, chewing *paan*, tobacco)	47.0	42.4	43.8
Total	164	354	518

Source: Fieldwork, 2001.

Note
Chi-square significant level ***1%; **5%; *10%.

wives of non-migrant men (4.2). After adjusting for the effect of age, the wives of non-migrants had 0.4 more children than the left-behind wives. A lower proportion of the left-behind wives had experienced an abortion (14 per cent), compared to a higher proportion of the wives of non-migrant men (21 per cent).

A little over three quarters of the left-behind wives lived in a joint family, compared to less than 50 per cent of non-migrant wives.[4] A higher proportion of the left-behind wives (26 per cent) were literate, compared to the non-migrant wives (19 per cent). The work participation rate (current work status) was higher for non-migrant wives. A significantly higher proportion of non-migrant wives worked for wages (43 per cent) compared to the left-behind wives (29 per cent). The use of contraception (sterilization and IUD, which may also cause RTI) was relatively low among left-behind wives (10 per cent) compared to the wives of non-migrant men (21 per cent).

Forty-two per cent of the non-migrant wives and 33 per cent of left-behind wives reported that their husbands consumed alcohol quite regularly. Fifteen per cent of the wives of non-migrant men and 33 per cent of the left-behind wives knew about, or suspected, their husbands of extramarital relations. The differences in the drinking habits and extramarital relations of husbands in the two groups were statistically significant. Consumption of alcohol may trigger extramarital sexual relations or divert money required for necessary health care.

Women's attitudes towards extramarital relations

Some of the male migrants who live in big cities far away from their wives visit commercial sex workers or develop relations with other women and raise parallel families in the city. Several studies have shown this type of behavioural change in men (Decosas *et al.* 1995; Harris 1993; Mishra 2002). Sometimes the left-behind wives of migrant men in rural areas do not even know of such relationships. Even if they did know, they could not do anything about it. They accept male superiority and the rights of men. Of men's extramarital relations, one woman said: 'If a man has extramarital relations with several women, he is still considered pure, but if a woman has extra-marital relations with only one person, she becomes impure' (*mard 17 handi dhund ke aai to bhi paak rahi par mehraru ek handi bhi dhundi to paak na rahi*). Many women say: 'Men have the right to do anything, like indulging in extramarital relations. They can pay and enjoy their lives.' Some women think that it is useless to stop a man from indulging in extramarital relations, since men cannot be controlled: 'You can tie four-legged creatures – animals – but you cannot tie two-legged creatures – men' (*Chargodawa bandhala, do godawa na bandhala*). On the question of fulfilling a woman's sexual desires in the absence of her husband, the reply was: 'If you want, you can control your desires' (*apana man aur dil jaise rakho waise rahi*). These statements show different social norms are for men and women. Women are taught to control their sexual desires, and if they cannot do so they face societal ostracism.

In this study a question about husbands' extramarital relations was asked of all the women. Eighty-five per cent of the non-migrant, and 67 per cent of the left-behind wives, denied any possibility of their husbands having extramarital affairs (see, Figure 11.4). The rest either suspected, or were sure, about the extramarital relations of their husbands. Nevertheless, a significantly higher proportion of the left-behind wives (13 per cent), compared to non-migrant wives (8 per cent), confirmed that they knew that their husbands had extramarital relations. However, some women (21 per cent of left-behind wives and 7 per cent of non-migrant wives) were not sure about the adultery committed by their husbands. According to them, a woman should not be surprised if she learns about the extramarital relations of her husband – this is the nature of a man.

General morbidity

Table 11.4 reveals that in both groups of women, general health symptoms were alike. Weakness was reported by 70 per cent of the respondents, followed by dizziness and backache (63 per cent each). The majority also reported having frequent headaches and back pain. The complaint of

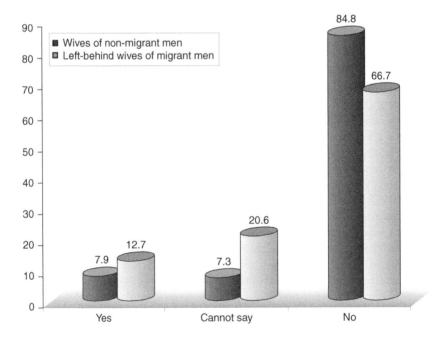

Figure 11.4 Wives' perception about extramarital relations of their husbands.

Table 11.4 Percentage of women reporting the symptoms of general health problems during the six months preceding the survey

Symptoms of general morbidity[a]	Wives of non-migrant men (%)	Wives left behind by migrant men (%)	Total women (%)
Headache	47.6	57.2	52.1
Back pain	62.8	62.4	62.5
Burning in hands and feet	36.6	27.1	30.1
Itching body	5.5	6.2	6.0
Dizziness	62.2	63.6	63.1
Weakness	70.1	70.1	70.1
Other	18.3	18.6	18.5
Total	164	354	518

Source: Fieldwork, 2001.

Note
a Multiple response.

headaches was significantly higher among left-behind wives, whereas a burning sensation in the hands and feet was significantly more common among the wives of non-migrant men. Six per cent of the women reported itching in the body, and 19 per cent other problems. Overall these problems are closely related to poverty, workload, lack of nutrition and mental stress (Sharma and Shivea 2000).

Symptoms of RTI/STI

In developing countries, mortality and morbidity due to reproductive tract infections and sexually transmitted infections are very high relative to other health problems. Because of the stigmatization of reproductive morbidity, women are often reluctant to report their problems, but every effort was made in this study to elicit accurate information from them. The respondents were assured of the confidentiality of the information provided.

Table 11.5 shows the prevalence of particular symptoms of reproductive morbidity in both the migrant and non-migrant groups, and Table 11.6 depicts the proportion of women who had experienced an abnormal vaginal discharge during the past six months, and the nature of the discharge. Abnormal vaginal discharge is regarded as one of the key symptoms of lower reproductive tract infection, whereas lower abdominal pain and lower backache are considered symptoms of cervicitis and/or pelvic inflammatory diseases (Garg *et al.* 2001). A thick, white discharge, irritation and itching are associated with candidiasis (vaginal infection), whereas a profuse discharge, burning urination, bad smell and lower abdominal pain are symptoms of trichomoniasis. Overall, in both the groups combined, burning or pain during

Table 11.5 Percentage of women reporting the symptoms of reproductive tract infection or sexually transmitted infection during the six months preceding the survey

Symptoms of RTIs/STIs[a]	Wives of non-migrant men (%)	Wives left behind by migrant men (%)	Total women (%)
Burning or pain or difficulty during urination	31.7	31.1	31.3
Itching or irritation around vagina	20.7	18.1	18.9
Sores on genitals	13.4	14.1	13.9
Pain during intercourse	18.3	16.9	17.4
Problems in menstruation[b]	27.0	26.4	26.6
Genital prolapse	6.1	4.0	4.6
Abnormal vaginal discharge	23.8	24.6	24.3
Total	164	354	518

Source: Fieldwork, 2001.

Notes
a Multiple response.
b Applicable to those who are not pregnant and who did not experience hysterectomy. Menstrual complications include irregular periods, scanty flow, fishy smell and severe back/abdominal pain.

urination was reported by 31 per cent of the women. Approximately a quarter of the respondents (both groups combined) reported problems in menstruation (27 per cent) and abnormal vaginal discharge (24 per cent). There was hardly any difference in the proportion of women who reported these problems by the migration status of their husbands.

Among those who reported vaginal discharge, 43 per cent said it was 'normal' since it occurred during the pre- and post-menstruation periods (Table 11.6). Nearly one third of the women complained about a 'thin, creamish discharge with a foul smell', and 20 per cent reported a 'thick, curdy' discharge. Nearly 50 per cent of the respondents who had abnormal discharge associated it with either itching or ulcer, or both. Eighty per cent of those who had experienced abnormal discharge associated it with either light fever or abdominal pain/backache, or both. Contrary to expectations, a higher proportion of non-migrant wives, compared to left-behind wives, complained of both itching and ulcers, and backache and fever associated with abnormal vaginal discharge. Among the other symptoms of RTI/STI, 19 per cent of all the women had experienced itching or irritation around the vagina, 17 per cent had experienced pain during intercourse, and 14 per cent had sores on their genitals (Table 11.5).

Considering all the symptoms of sexually transmitted infections, it may be said that 9 per cent of the non-migrant wives and 5 per cent of the left-

Table 11.6 Percentage of women reporting vaginal discharge and its associated
nature and characteristics during the six months preceding the survey

Nature and characteristics of vaginal discharge[a]	Wives of non-migrant men (%)	Left-behind wives of migrant men (%)	All wives (%)
Experienced vaginal discharge preceding six months of the survey			
Not experienced	60.4	55.9	57.3
Experienced	23.2	24.6	24.1
Currently experienced	16.5	19.5	18.5
Total	164	354	518
Texture of vaginal discharge			
Normal (before or after menstruation)	40.0	44.2	43.0
Thick, curdy white	24.6	17.9	19.9
Thin, creamish with foul smell	27.7	32.1	30.8
Thick, grey with foul smell	7.7	5.8	6.3
Total	65	156	221
Vaginal discharge associated with[b]			
Itching	20.5	14.2	16.7
Ulcer	5.1	5.7	4.0
Both	28.2	24.1	25.4
None	46.2	57.5	54.0
Total	39	87	126
Vaginal discharge associated with[b]			
Abdominal or back pain	17.9	18.4	18.3
Fever	7.7	5.7	6.3
Both	59.0	54.0	55.6
None	15.4	21.8	19.8
Total	39	87	126

Source: Fieldwork, 2001.

Notes
a Multiple response.
b Refers only to those wives who experience abnormal discharge.

behind wives were suffering from symptoms of candida (thick, white discharge, irritation and itching), and 12 per cent of the non-migrant wives
and 16 per cent of the left-behind wives were suffering from the symptoms
of trichomoniasis (profuse discharge, burning urination, bad smell and lower
abdominal pain). However, these results should be interpreted with caution
as the reports of these symptoms were not clinically verified.

Overall, no significant difference was observed in the prevalence of a
particular morbidity or combination of morbidities between the two study
groups. More than a quarter of the left-behind wives did not communicate

these problems to anyone, probably because of the culture of silence and the stigma attached to these morbidities. Indian women suffer silently from sexually transmitted infections, and thereby increase their vulnerability to HIV infection. The fear of ostracism and rejection makes many women hide or neglect their condition, even if they know that the only source of infection is their husband (Saran 2002). For instance, a woman suffering from a white discharge may disclose her problem to her husband, who, instead of taking her to the health centre for treatment, may make allegations against her as having had relations with other men, and insist that this is the cause of her discharge. This is one of the reasons why women do not dare disclose their reproductive illnesses.

Health-seeking behaviour

In this study women who reported health problems were asked who they consulted about the problem. A large majority of the women did not seek any treatment for their reproductive tract infections (65 per cent did not seek treatment for abnormal vaginal discharge, and 80 per cent did not seek treatment for menstrual problems). The main reason reported by them for not seeking treatment was non-affordability of the cost of treatment. Those who sought medical advice did not consult government doctors or auxilliary nurses/midwives because the government system does not function properly, forcing poor people to pay for services that are supposed to be free or subsidized. For treatment of discharges, approximately 6 per cent of the women consulted traditional healers. Many women did not seek treatment because the doctor was male and they did not want to discuss 'dirty' or 'shameful' problems with a man. Others did not perceive the problems as serious; they thought they were 'normal' problems associated with a woman's life.

Regarding the irregularity of menstruation (irregularity of menstruation has not been referred to before as a symptom of RTI or STI), there is a very common perception that if a woman has already given birth to children there is no need for her to regulate her menstrual cycle. Rather, irregular periods are considered good for protection against unwanted pregnancies. To treat their reproductive health problems, most of the women depend on home remedies. They may often use a 'herbal paste', or wash their sexual organs with 'rice water' (left-over water after washing the rice grains), *neem* water (*neem* leaves are soaked in water), or diluted 'Dettol'. Some also take an *ayurvedic* health tonic known as *Ashoka Risht*.

Prevalence of reproductive tract infections

In the community-based studies undertaken in India (Bang *et al.* 1989; Bhatia and Cleland 1995; Shenoy *et al.* 1997), the range of self-reported morbidity varies from between 40 to 84 per cent. The present study shows that 88 per cent of non-migrant wives and 94 per cent of left-behind wives had reported at least one symptom of RTI/STI (burning sensation during urination, itching, sores on the genitals, pain during intercourse, problems in menstruation and abnormal menstrual discharge), along with other gynaecological morbidities such as weakness, backache and dizziness. But weakness, dizziness and backache may also be caused by malnutrition or heavy workloads; therefore, even after excluding these symptoms, a significantly higher proportion of the left-behind wives (65 per cent) compared to non-migrant wives (55 per cent) reported one or other symptom of RTI/STI.

Reproductive morbidities are determined mainly by menstrual hygiene, the risk behaviour of one of the partners, contraceptive infection, the socio-economic characteristics of the respondents and health seeking behaviour. Both the study group and the reference group were found to be homogeneous in terms of socio-economic characteristics (caste and land), and such factors as sanitation and hygiene.

Table 11.7 presents the prevalence rate of at least one symptom of

Table 11.7 Prevalence of any reproductive morbidity among women according to their background characteristics and husband's perceived behaviour

Background characteristics	Wives of non-migrants (%)	Left-behind wives of migrants (%)	Total wives (%)
Standard of living (SLI)			
Low	64.3	58.5	60.5
Medium	58.6	62.9	61.4
High	34.2	75.0	64.4
Type of family			
With in-laws (joint)	50.6	67.2	63.5
Without in-laws (nuclear)	58.6	59.3	59.0
Age of respondent			
<25 years	57.5	65.7	63.8
>25 years	54.0	65.0	61.0
Literacy			
Illiterate	54.9	64.6	63.8
Literate	54.8	67.0	61.0
Marital duration			
Less than ten years	60.0	66.2	64.7
Ten to twenty years	53.8	61.5	58.7
More than twenty-one years	51.3	71.0	63.9
Children ever born (CEB)			
≤3 CEB	59.4	66.7	65.0
≥4 CEB	52.0	63.0	58.3

continued

Table 11.7 continued

Background characteristics	Wives of non-migrants (%)	Left-behind wives of migrants (%)	Total wives (%)
Experienced abortion			
No	55.0	64.7	61.8
Yes	54.3	68.6	62.8
Work in own or others' field			
No	57.5	65.4	63.4
Yes	52.7	65.0	60.2
Work for wages			
No	50.0	65.2	61.0
Yes	61.4	65.4	63.8
Ever used contraception			
No	56.8	63.5	61.5
Yes	50.0	72.2	63.6
Ever used (IUD or sterilization)			
No	58.8	63.8	62.2
Yes	41.2	79.4	60.3
Drinking habit of husband			
No	53.7	62.8	59.2
Yes	56.5	70.4	69.2
Smoking habit of husband			
No	55.0	64.0	61.1
Yes	54.5	68.3	64.2
Suspect extra-marital relations of husband			
No	51.8	63.6	59.2
Yes	72.0	68.6	69.2
Any other habit in husband (gambling, chewing paan, *tobacco)*			
No	49.4	62.3	58.4
Yes	61.0	69.3	66.5
Duration of migration			
Up to five years	–	68.0	–
Five to ten years	–	63.9	–
More than ten years	–	63.7	–
Time of migration			
Before marriage	–	63.3	–
Less than five years after marriage	–	63.4	–
Six to ten years after marriage	–	83.3	–
More than ten years after marriage	–	62.2	–

Table 11.7 continued

Background characteristics	Wives of non-migrants (%)	Left-behind wives of migrants (%)	Total wives (%)
Destination[a]			
Western states	–	72.5	–
Northwestern States	–	61.4	–
Adjoining states	–	61.3	–
Other states	–	70.0	–
Occupation of migrants[b]			
Sale worker	–	75.0	–
Services	–	67.6	–
Production + labourer	–	62.8	–
Construction	–	61.9	–
Textiles	–	66.7	–
Transportation	–	70.0	–
Defence	–	72.7	–
Other	–	75.0	–
Do not know	–	56.7	–
Total	164 (54.9)	354 (65.3)	518 (62.2)

Source: Fieldwork, 2001.
Notes
a Western states (Maharashtra, Gujarat, Madhya Pradesh, Rajasthan)
   Northwestern states (Punjab, Haryana, Himachal Pradesh, J&K, Delhi Metro Region, Uttaranchal)
   Adjoining states (Jharkhand, Chattisgrah, Orissa, Uttar Pradesh)
   Other states (North Eastern and Southern states).
b Sales worker (vender, milkman, helper in grocery shop)
   Production and labourer (industrial labourer (paper, iron scrap), loader, fitter, welder)
   Textile (labourer, power loom, tailoring, carpet-making)
   Transport (truck or bus driver, cleaner/conductor)
   Service (barber, watchman, washerman, sweeper)
   Other (agricultural labourer, clerk, traditional musician and dancer, teacher).

RTI/STI according to socio-economic and demographic characteristics and perceived life-style of the husband. By and large, the reporting of morbidity was higher in almost all the subgroups of the left-behind wives compared to the non-migrant wives. In both groups of women – non-migrant and left-behind wives – the prevalence of self-reported morbidity was higher for those women who had lower parity, whose husbands drank and smoked, had extramarital relations and other habits such as gambling, and chewing tobacco or *paan masala*. Left-behind wives differed from non-migrant wives in terms of the Standard of Living Index[5] (SLI), type of family, marital duration, working for wages and use of contraceptive methods. Among the wives of non-migrant men, the prevalence of RTIs/STIs was higher among those who belonged to the lower SLI category, lived in a nuclear family and worked for wages. On the other hand, the perceived symptoms of RTIs/STIs were higher among those who had a better standard of living, lived in a joint family and had used any contraceptive method (sterilization and IUD).

Determinants of reproductive morbidity

To determine the effect of different variables on reproductive morbidity, a logistic regression analysis was carried out wherein the prevalence of reproductive morbidity was taken as the dependent variable, and standard of living, marital duration, age of the respondent, use of contraception, the respondent's perception of her husband's extramarital relations, consumption of alcohol, habits such as gambling, and chewing *paan masala* and tobacco were taken as predictor variables (Table 11.8).

Left-behind wives were more likely to report morbidity than were the wives of non-migrant men. The odds ratio of reporting to not reporting morbidity was 1.4 times higher for left-behind wives than for non-migrant wives. Of all the women (left-behind and non-migrant wives) who knew, or suspected, the extramarital relations of their husbands, those whose husbands had habits such as gambling and/or chewing tobacco or *paan masala* were more likely to report reproductive morbidity, compared to those whose husbands did not have such habits. With respect to marital duration, women of longer marital duration (more than ten years) were less likely to report symptoms of RTIs/STIs than were women of shorter marital durations (less than ten years). For the wives of non-migrant men, it was found that, after controlling for the effects of other variables, having extramarital relations significantly influenced the prevalence of RTIs/STIs in a woman, with an odds ratio of 3.1. In other words, those women who reported that their husbands had extramarital relations were more likely to suffer from RTIs/STIs than were those women whose husbands reportedly did not have such relations.

For the left-behind wives, the results of logistic regression revealed that those women who belonged to a better SLI group were more likely to perceive reproductive morbidity than were women who belonged to a lower group on the Standard of Living Index. Women whose husbands had migrated to the western states such as Maharashtra, Gujarat, Rajasthan and Madhya Pradesh were also more likely to report reproductive morbidity, compared to women whose husbands had migrated to other parts of the country. Further, women whose husbands migrated after a few years of marriage (five or more) were significantly more likely to report the prevalence of RTIs/STIs than were those women whose husbands had migrated before marriage. The likelihood of reporting reproductive morbidity was reduced as marital duration increased. In the longer marriage duration category, perceptions about the extramarital relations of the husband did not have a significant effect on the prevalence of RTI/STI symptoms. However, it is possible that such behaviour among migrant men would be under-reported by their wives due to their faith in their husbands.

Table 11.8 Logistic regression showing the effect of various socio-economic and behavioural characteristics on prevalence of any reproductive morbidity

Predictor variables	Total women	Wives of non-migrant men	Wives left behind by migrant men
	Exp(B) (SE)	Exp(B) (SE)	Exp(B) (SE)
Migration of husband			
Women of non-migrant (R)			
Wives left behind by migrant men	1.420*	–	–
	(0.205)	–	–
Standard of living			
Low SLI (R)			
High SLI	1.100	0.552	1.714**
	(0.202)	(0.367)	(0.257)
Extramarital relations of husband			
No extramarital relations (R)			
Any extramarital relations	1.534*	3.084**	1.403
	(0.236)	(0.542)	(0.277)
Drinking habit of husband			
Does not drink (R)			
Husband drinks	1.179	0.953	1.408
	(0.229)	(0.382)	(0.301)
Smoking habit of husband			
Does not smoke (R)			
Husband smokes	0.936	0.668	0.961
	(0.243)	(0.449)	(0.309)
Other bad habits of husband (gambling, chewing paan, *tobacco)*			
No other bad habit (R)			
Any other habit	1.536**	1.634	1.491
	(0.194)	(0.343)	(0.247)
Marital duration			
Less than ten years (R)			
Ten to 20 years	0.427**	0.537	0.342***
	(0.351)	(0.848)	(0.409)
More than twenty years	0.407**	0.483	0.277**
	(0.421)	(0.933)	(0.520)
Contraceptive methods (sterilization or IUD)			
Never used (R)			
Ever used	1.169	0.698	1.569
	(0.225)	(0.373)	(0.307)

continued

Table 11.8 continued

Predictor variables	Total women	Wives of non-migrant men	Wives left behind by migrant men
	Exp(B) (SE)	Exp(B) (SE)	Exp(B) (SE)
Experience of abortion			
Never experienced (R)			
Ever experienced	1.074 (0.165)	1.028 (0.253)	1.003 (0.234)
Age of respondent[a]	1.019 (0.022)	1.002 (0.038)	1.041 (0.030)
Number of visits of migrant to the place of origin in a year[a]	–	–	0.701 (0.249)
Destination of migrant			
Other states (R)			
Western states	–	–	1.590*
	–	–	(0.271)
Time of migration			
Before marriage (R)			
Within five years of marriage	–	–	1.121 (0.269)
After five years of marriage	–	–	1.834* (0.360)
Constant	1.050 (0.539)	2.812 (1.017)	0.688 (0.718)

Source: Fieldwork, 2001.

Notes
*** $p < 0.01$;
** $p < 0.05$;
* $p < 0.1$.
a Continuous variable.
Figure in parenthesis presents standard error.

In short, for left-behind wives, SLI, marital duration and migration factors (time of migration and migration destination) play a significant role in the prevalence of any RTI/STI. Women whose husbands had migrated after five years of marriage had a higher prevalence of reproductive morbidity. One plausible explanation for this phenomenon could be that these migrant men had experienced such sexual satisfaction in their married life that it was difficult for them to forgo sexual relations when they were away from their wives. Regarding the destination of migrants, it is suggested that the prevalent culture in western states, particularly in Maharashtra and

Gujarat, is more liberal, where exposure to risk is higher and, therefore, the chances of contracting infections are also higher. The findings of the National AIDS Control Organization also show that the prevalence of these infections is higher in metropolitan areas and in the industrially developed states of Maharashtra and Gujarat (UNAIDS 2002).

The link between migration and sexually transmitted infections: two case studies

The following case studies illustrate how male migrants can spread sexually transmitted infections to their unsuspecting wives.

Woman suffering from cancer of the uterus

Mahua Devi (45) is an illiterate scheduled caste woman suffering from cancer of the uterus.[6] She married at the age of 16. Of her eight conceptions, one male and three female children survive. Before marriage, her husband had migrated to Calcutta. After marriage, he did not take his wife with him to Calcutta, but kept visiting home regularly. Other than general health problems, Mahua suffers from gynaecological problems, such as a thick, white discharge, irregular menstruation, bleeding during intercourse and a burning sensation while urinating. She also suffers from a vaginal ulcer and swelling in the abdomen. When her problem became acute she visited a private clinic where she was told that she had cancer of the uterus and would survive for only another seven or eight months. She attributes her health problems to her husband's migration and his sexual behaviour when he is away from her: 'He used to visit several women in Calcutta. I had suspicions of contracting a sexually transmitted disease from my husband when I first developed vaginal itching, which has taken this form now. I am paying for the misdeeds of my husband.' She feels that such a disease can be contracted by any woman whose husband has migrated and developed extramarital relations. In her opinion the husband and wife should stay together.

Husband infected with HIV/AIDS

Phulmati Devi (30) is the mother of three children. She is a landless woman from a lower caste. She completed tenth-grade education and teaches in a school under the informal education programme. Since this programme is not running well, she does not get paid on a regular basis. She also works as an agricultural labourer to supplement her income and eke out a modest existence for her and her children.

Phulmati's husband migrated after four years of marriage to work in a power loom in Mumbai. He earns Rs.70 (equivalent to about US$1.50) per day and irregularly remits some money to his family. Last year he sent a total of Rs.4,000, which Phulmati spent in purchasing consumables. She did not have enough money to repair her *kuchcha* house (thatched house made of mud and clay), which is falling apart in such a way that it is very difficult to enter the house. It is very risky to live in such a place.

Her husband does not send money home regularly. Moreover, these days he is sick; therefore most of his earnings are spent on his own treatment. His doctor has told him that he is suffering from AIDS. Phulmati had never heard of AIDS before this and her husband blames her for his disease. But she asserts that she never had extramarital relations with anyone. Apart from her tight economic situation she is worried about her husband's suspicion of her character, and of acquiring AIDS herself.

The first case study suggests how the reproductive health of a woman can be affected when her husband indulges in extramarital relations. In the second case, the wife is suffering from mental stress as the husband has blamed her for his illness. Moreover, she may also contract the infection from her husband. These case studies show that although the decision to migrate is good for the survival of the family, sometimes it may become a fatal health risk if migrant men are not careful about their sexual behaviour.

Knowledge about RTI/STI and HIV/AIDS

In 1999, the Prime Minister of India emphasized 'maximizing public knowledge about HIV/AIDS since there is no cure for the disease. Prevention and prevention alone, is the cure. It is one of the four fronts to battle against HIV/AIDS' (Kumar 2001: 40). In the current study questions were asked of all the women regarding their awareness about RTIs/STIs and HIV/AIDS. If they had heard of these problems they were asked about the source of their knowledge and the mode of cure.

The results from the study show that 89 to 94 per cent of the women in different groups have heard about RTI (the infection that occurs in the genitalia). Of those who have heard of RTI, the source of information was relatives and friends (Table 11.9). When they were asked about their knowledge regarding the cause of RTI, more than 45 per cent said that they did not know the cause. Of those who did know the cause, 41 per cent said it was due to 'heat in the stomach', 24 per cent said it was due to a lack of personal hygiene, and 5 per cent said it was due to sex during menstruation. The respondents perceived that heat in the stomach or blood is caused by taking oily and spicy food, chillies and jackfruit. Many women, who named unhygienic conditions as the cause of RTI, defined an unhygienic environment as urinating in the same place where someone else had urinated before and the place was still dirty. More than three-quarters of the women considered RTIs to be curable.

Sexually transmitted infection is known in the community as *lasora bimari* (sticky disease). The respondents were asked the question: 'Have you ever heard of a disease that can be transmitted from a man to a woman or vice versa after sexual intercourse?' Nearly half (48 per cent) of the women reported that they had heard about such an infection, and that it was caused by sexual intercourse. Regarding the source of their information, the majority said that they had learned about the disease from relatives and friends (84 to 91 per cent). Only about 11 per cent of the women said they had learned about the disease from their husbands, and 7 per cent said they had heard about it from the radio. Three-quarters said that the disease was caused by having multiple sex partners. Nineteen per cent of the women said that they did not know exactly how the disease was caused. More than 62 per cent of the women said that STI was curable. There was no significant difference between the wives of non-migrant wives and left-behind wives regarding their awareness of RTIs/STIs. The reason could be that their main source of information is friends and relatives.

Women's awareness of AIDS and its causes was obtained in the same manner as that of RTI/STI. Only 6 to 11 per cent of the respondents (eleven non-migrant wives and thirty-eight left-behind wives) had heard the term 'AIDS'. The majority said that they had learned about AIDS from the radio. Of forty-nine women who had heard of AIDS, only one woman had correct knowledge about the transmission of the disease.

Limitations

There are a number of limitations to this study. First, the information collected on general and reproductive health is based entirely on self-reported symptoms, and no clinical examination was conducted to verify the presence of disease. Second, Indian society generally follows the monogamous form of marriage. A woman is not supposed to have sexual relations with any man other than her husband. Although there is always a chance that some women may break this norm, particularly when their husbands are away, it is not an accepted social norm. In this context it would be appropriate to assume that if a woman has contracted an STI it is from her husband. The third limitation of the study is that only women were included in the sample; their husbands were not interviewed. Although it would have been ideal to interview the husbands it was not possible to track them, many of whom lived hundreds of kilometres away.

Summary and conclusions

In India, rural to urban migration is male-dominated and mainly a survival response to the prevailing poverty in rural areas. Because of the costlier life in urban areas, such migrants are forced to leave their families behind. Once at their destination many indulge in unsafe sexual behaviour and some

Table 11.9 Awareness about RTIs/STIs and AIDS, source of information and mode of transmission among respondents

	Awareness about RTIs		Awareness about STIs		Awareness about AIDS	
	Wives of non-migrant men % (No.)	Wives left behind by migrant men % (No.)	Wives of non-migrant men % (No.)	Wives left behind by migrant men % (No.)	Wives of non-migrant men % (No.)	Wives left behind by migrant men % (No.)
Have you heard about such infection as[a]						
Yes	87.2	94.4	50.0	50.0	5.7	10.7
Total	164	354	164	354	164	354
Source of information[a]						
Radio	2.8 (4)	6.6 (22)	8.5 (7)	11.3 (20)	81.8 (9)	78.9
Doctor	0.7 (1)	1.5 (5)	1.2 (1)	3.4 (6)	15.8 (6)	12.2 (6)
Teacher	–	0.3 (1)	–	1.1 (2)	–	–
Friends and relatives	9.3.0	91.9	91.5	84.2	27.3 (3)	26.3 (10)
Husband	2.5 (5)	3 (10)	12.2 (10)	16.4	36.4 (4)	10.5 (4)
Others	20.3	23.7	–	–	–	–
Total	143	334	83	177	11	38

Mode of transmission[a]

Multiple sex	7.6 (13)	6.6 (22)	78.0	75.7	63.6 (7)	63.2 (24)
Needle/blade	–	–	–	–	36.4 (4)	34.2 (13)
Mother to child	0.6 (1)	1.5 (5)	2.4 (2)	5.1 (9)	45.5 (5)	34.2 (13)
Lack of personal hygiene	24.7	23.7	2.4 (2)	1.1 (2)	–	–
Sex during periods	5.3 (9)	4.2 (14)	–	–	–	–
Heat in stomach	41.2	40.7	–	27.3 (3)	28.9 (11)	–
Blood transfusion	–	–	1.2 (1)	0.6 (1)	9.1 (1)	–
Other	6.5 (11)	8.7	–			–
Do not know	45.3	45.5	19.5 (16)	19.2	36.4 (4)	34.2 (13)
Total	143	334	83	177	11	38
Curable[a]						
Yes	80.4	75.7	69.9	62.1	(2)	18.4 (7)
No	2.8 (4)	6.0	1.2 (1)	7.3 (13)	(6)	39.5 (13)
Do not know	16.8	18.3	29.0	30.5	(3)	42.1 (16)
Total	143	334	83	177	11	38

Source: Fieldwork (2001)

Note
a Multiple responses.

contract STIs. Many studies have noted that HIV is spreading first to high-risk groups and then to other people. Migrants play a significant role in this process. It is not migration per se, but the behaviour of the migrants that plays a greater role in spreading the infection.

In the context of rural Bihar, this study explored the reproductive health of the left-behind wives of migrant men with the help of self-reported symptoms. It was found that the overall prevalence of reproductive infections was high among rural women, and that after controlling for the effects of other variables, left-behind wives had a significantly higher level of reproductive morbidity than the wives of non-migrant men. For the latter, the husbands extramarital relations emerged as a significant factor in determining RTIs/STIs, but, in the case of left-behind wives, the other significant determinants of reproductive morbidity were a lower standard of living, shorter marital duration, husbands' migration to western states, and the timing of the husbands' migration. The husbands' extramarital relations did not play a significant role in determining the reproductive morbidity of left-behind wives.

Thus the findings lead to another set of questions: Why is reproductive morbidity high among left-behind wives? Is it that they do not treat themselves if they develop reproductive morbidity? Or are these women unable to take care of their reproductive morbidity in the absence of their husbands, either because of an increase in workload, or because there is no one to take them to a doctor or they feel reluctant to discuss such matters with others? These questions need further investigation.

However, taking both groups together (left-behind wives, as well as non-migrant wives), the extramarital relations of the husband emerge as a significant determinant of women's reproductive morbidity. As far as knowledge of RTI/STI and HIV/AIDS is concerned, a high proportion of women have heard about an RTI/STI, but their knowledge is incorrect and incomplete. The reason for this could be that they get information mainly from informal sources (friends and relatives) who themselves might not have complete knowledge about the diseases. Knowledge about HIV/AIDS is very limited among rural women.

The best way of risk reduction could be the sustainable economic development of the rural areas in order to reduce distressed male out-migration. Apart from that there is a need to provide information to change the risk behaviour of people such that they practise safe sex and abstinence. To reduce reproductive morbidity among women, men should follow the ABC of sex: Abstain from sex, Be faithful to their wives, or use Condoms. Accurate knowledge about sexually transmitted diseases should be disseminated via community radio, street plays (*nautankies/tamasha* – important sources of information in the region), pamphlet distribution, poster campaigns, and propaganda during *melas* (fairs and festivals).

Since medical services are not easily accessible to everyone in the villages, the foremost suggestion would be that along with government service providers, private medical practitioners, teachers, traditional healers and untrained *dais* should also be trained in sanitation and health issues related to RTI/STI and HIV/AIDS transmission. This type of training could include informing people about bodily functions, sanitation, the harmful effects of intoxicants, and safe sexual practices.

Acknowledgements

We acknowledge the Ford Foundation Grant received at the International Institute for Population Sciences under the project titled 'Strengthening the Teaching and Research Activities in Reproductive Health at IIPS (1996–2000)' by Grant No. 950–1006 for undertaking this research.

Notes

1 For the present study the term 'wives left behind by migrant men' is used for wives whose husbands have migrated to other states for the purpose of employment, and 'wives of non-migrant men' or 'wives of non-migrants' for those women whose husbands have never migrated.
2 Refers to proportion of male workers who are employed but seek, or are available for, additional work.
3 Marginal households are those that have less than one acre of land.
4 *Encyclopedia Britannica* (2004) defines a joint family as a 'Family in which members of a descent group live together with their spouses and offspring in one homestead and under the authority of one of the members'.
5 The Standard of Living Index is computed by taking into account possession of household amenities, type of house and landholding size.
6 All names are pseudonyms.

References

Armstrong, S. (1995) 'AIDS and migrant laborers', *Populi*, 22, 12 (December 1994 to January 1995): 13.
Bang, R., Bang, A., Baitule, M., Chaudhary, Y., Sarmukaddam, S. and Tale, O. (1989) 'High prevalence of gynaecological disease in rural Indian women', *Lancet*, 8629: 85–8.
Bhatia, J.C. and Cleland, J. (1995) 'Self reported symptoms of gynecological morbidity and their treatment in south India', *Studies in Family Planning*, 26, 4: 203–13.
Carlier, J-Y. (1999) *The Free Movement of Persons Living with HIV/AIDS*, Brussels: HIV/AIDS Programme of the European Union.
Cruz, P. and Azarcon, D. (2000) 'Filipinos and AIDS: it could happen to you', in *Conveying Concerns: Media Coverage of Women and HIV/AIDS*, Washington, DC: Population Reference Bureau, pp. 20–21.
De Schryver, A. and Metheus, A. (1991) 'Sexually transmitted diseases and migration', *International Migration*, 29, 1: 13–28.
Decosas, J., Kane, F., Anafri, J.K., Sodji, K.G. and Wagner, H.U. (1995) 'Migration and AIDS', *Lancet*, 346, 8978: 826–8.
Encyclopedia Britannica (2004) 'Joint family', Encyclopedia Britannica Premium

Service. Online, available at: www.britannica.com/eb/article?eu=44900 (accessed 11 September 2004).

Garg, S., Bhalla, P., Sharma, N., Sahay, R., Puri, A., Saha, R., Sodhani, P., Murthy, N.S. and Mehra, M. (2001) 'Comparison of self-reported symptoms of gynaeco-logical morbidity with clinical and laboratory diagnosis in a New Delhi slum', *Asia-Pacific Population Journal*, 16, 2: 75–92.

Gilbert, T.J. (1997) 'The HIV/AIDS epidemic in sub-Saharan Africa', *Population Bulletin*, Washington, DC: Population Reference Bureau, pp. 28–35.

Government of India (1999) *Reproductive and Child Health Project-Rapid Household Survey (Phase-I & II) 1998–1999*, Mumbai: International Institute for Population Sciences.

Government of India (2001) *Employment and Unemployment in India 1999–2000*, National Sample Survey No. 458, 55th Round, New Delhi: NSSO.

Gupta, K. and Singh, S.K. (2002) 'Social networking, knowledge of HIV/AIDS and risk-taking behavior among migrant workers', Paper Presented at the Regional Conference of IUSSP, Bangkok, 11–13 June.

Haour-Knipe, M. (2000) *Migration and HIV/AIDS in Europe*, Geneva: International Organization for Migration (IOM) and the University Institute for Social and Pre-ventive Medicine, p. 4.

Harris, J.C. (1993) 'Eco-earth in rural environment', *Health for the Millions*, 1, 2: 3–4.

International Institute for Population Sciences and ORC Macro (2001a) *National Family Health Survey (NFHS-2), Bihar 1998–99*, Mumbai: IIPS.

International Institute for Population Sciences and ORC Macro, (2001b) *National Family Health Survey (NFHS-2), India 1998–99*, Mumbai: IIPS.

Inversen, A., Fugger, L., Eugen-Olsen, J., Balslev, U., Jensen, T., Wahl, S., Ger-stoft, J., Mullins, J. and Skinhoj, P. (1998) 'Cervical human immuno deficiency virus type 1 shedding is associated with genital B-Chemokine secretion', *Journal of Infectious Diseases*, 178, 5: 1334–42.

Kumar, A. (2001) 'Combating AIDS in rural areas', *Kurukshetra* (July): 40–41.

Mishra, A. (2002) 'Male sexual behaviour and reproductive health with family and away from family in Delhi', unpublished Ph.D. thesis, International Institute for Population Sciences, Mumbai.

Nanda, A.R. (2002) 'Not just a number game', *Seminar: Web Edition.* Online, avail-able at: hdrc.undp.org.in/resources/gnrl/ThmticResrce/PopnDmgphy/A_R_ per cent20Nanda-2002.htm (accessed 16 September 2004).

Population Reference Bureau (2001) *Conveying Concerns: Media Coverage of Women and HIV/AIDS*, Washington, DC: Population Reference Bureau, p. 19.

Population Report (1993) *Controlling Sexually Transmitted Diseases*, 21, 1.

Population Report (1998) *People on Move: New Reproductive Health Focus*, 24, 3.

Population Report (2002) *Youth and HIV/AIDS: Can We Avoid Catastrophe?*, 29, 3.

Rasmussen, S., Eckmann, L., Quayle, A., Zhang, Y., Anderson, D., Fierer, J., Stephens, R. and Kagnoff, M. (1997) 'Secretion of proinflammatory cytokynes by epithelial cell in response to chlamydia infection suggests a central role of Epithe-lial cells in chlamydial pathogenesis', *Journal of Clinical Investigation*, 99, 1: 77–87.

Saran, S. (2002) 'Are you positive? Women at risk', in *Conveying Concerns: Media Coverage of Women and HIV/AIDS*, Washington: Population Reference Bureau, p. 9.

Sharma, G. and Shivea, M. (ed.) (2000) *National Profile on Women, Health and Devel-*

opment, *Country Profile – India*, New Delhi: Voluntary Health Association of India, pp. 156–67.

Shenoy, K.T., Shenoy, S. and Gopalkrishnan, K. (1997) 'Reproductive health and gynecological morbidity in Kerala', unpublished paper.

Theus, S., Harrich, D., Gaynor, R., Radolf, J. and Norgar, M. (1998) 'Treponema pallidum, lipoproteins, and systematic lipoprotein analoges induce human immunodeficiency virus type 1 gene expression in monocytes via nf-kb activation', *Journal of Infectious Diseases,* 177, Supplementary 3: 941–50.

Tidco-India Policies (2003) *National Health Policy 2002.* Online, available at: www.tidco.com/india_policies/other_sect_GoI_policies/healthpolicy.asp (accessed 16 September 2004).

Todaro, M.P. (1976) *Internal Migration in Developing Countries*, Geneva: International Labour Office.

UNAIDS and IOM (1998) 'Migration and AIDS', *International Migration,* 36, 4: 445–68.

UNAIDS (2002) 'Focus: AIDS and mobile populations', *Report on Global HIV/AIDS Epidemic,* Geneva: UNAIDS.

United Nations (1995) *Population and Development, Volume 1.* Programme of Action Adopted at the International Conference on Population and Development, Cairo, 5–13 September 1994, New York: United Nations, p. 30.

12 Some policy issues on migrant health

Gavin W. Jones

Introduction

This chapter pulls together some of the implications for health planning posed by the findings reported on in the earlier chapters. Migration – particularly within-country temporary or circular migration, and internationally, labour migration and unrecorded and 'illegal' migration – has always posed particular problems for economic and social planning. The same is true for health planning. Because of their temporary or 'illegal' status, labour migrants are placed in a lower priority situation in relation to services provided by the (often unwilling) 'host' government. This raises both ethical and political issues. On the former, it seems that many governments have avoided ratifying the 1990 UN Convention on the Protection of Migrant Workers and Members of their Families, thus avoiding the responsibility of providing certain services for labour migrants. On the latter, governments are usually under pressure from their electorate, or from whatever avenues of public expression of opinion exist, to give priority to their own citizens in the provision of services. Yet even if we accept this narrow view of the responsibilities of national governments, the health problems of migrants (documented or undocumented), if left unattended, or not fully dealt with by the existing health services, can also pose serious threats to the health of the host population. In China, for example, the SARS epidemic showed the government how vulnerable it was to the possibility of spread of the disease by the 'floating population', which was largely out of the reach of the urban health services.

Dealing with underlying and proximate issues of migrant health

There are different levels at which a particular health issue for migrants can be addressed. Take, for example, the issue of the need for 'illegal' labour migrants to access affordable medical care in the case of emergencies. This might be addressed at the proximate level by the intervention of non-government organizations (NGOs) providing health care for such migrants in places where they are known to be present in large numbers. At one remove from this 'coal-face' assistance is the need for employers to be penalized for knowingly employing undocumented migrants, and to be required to meet labour standards, including prevailing basic wage levels and provision of adequate medical facilities, for all migrant employees. At a further remove, there is the need for sending countries to adopt development strategies that work at the root causes of the poverty and political and economic instability that encourage undocumented migration. A holistic approach to the problem can no more ignore the need for 'coal-face' approaches that help to deal with the immediate problem of health emergencies of 'illegal' workers than it can ignore the need to deal with the root causes of undocumented migration.

Levels of policy intervention

Policy interventions with regard to migrants require actors, and it is important to be aware of the possible identity of these actors, and the different levels at which policy can be made and implemented with regard to migrants. One is the level of international agencies, which can pressure governments to adopt certain policies and conventions. At the level of government policy, in some cases it may be parliaments or cabinets enacting new laws or regulations; in others, planning agencies, government departments, quasi-government institutions, or local government agencies changing regulations or introducing new programmes. If we are advocating changes to policy, we need to know at what level of government action needs to be taken.

In general, government policy works best in 'regular' situations. Government policy and services typically do not adapt well to 'messy' situations, and governments are not good at taking nimble action in changing circumstances.[1] To deal with irregular or rapidly changing circumstances, it may often be preferable if the government does *not* have a declared policy, rather than having a clumsy policy or one that may attract opposition, and thus be an obstacle to achieving desired ends. Government encouragement of civil society or NGO activity may often be a better way to go, although ultimate responsibility for policy affecting migrant welfare must rest with the government.

There are many cases, though, where a better understanding of a problem may be all that is needed to get governments to change policy, or (to put it slightly less optimistically) at least to take the first necessary step towards policy change. This is where the role of good research, and advocacy based on it, comes in. Even the simple fact of the feminization of labour migration from countries such as the Philippines (60 per cent) or Indonesia (65 per cent: Hugo 1996) is not widely recognized by policy-makers. This is of some concern, because the greater vulnerability of women to abuse and sexual violence, particularly as 'illegal' migrants, has been well documented (Constable 1997; Cox 1997; Lim and Oishi 1996; Shah and Menon 1994).

The implications for physical and psychological health are serious. Different approaches in planning for migrant welfare may therefore be needed when women make up the majority of the migrants.

However, we cannot always expect the government's role to be benign; sometimes government policy may be an important *obstacle* to the health of migrants. Governments may, for example, seek actively to discourage actions that will increase labour costs of businesses, in the interests of raising competitiveness in world markets and maximizing employment (for examples of tight restrictions on labour union activity in Indonesia see Hadiz 1997; Manning 1998, ch 8). Similarly, in the interests of maximizing revenues from labour migrants, the Indonesian and Philippines governments have adopted rather 'non-interventionist' stances on the conditions of their labour migrants overseas, despite the many problems these workers face (see e.g. Kassim 1996).[2] On another level, government functionaries whose job it should be to protect the interests of migrants may be part of the problem. Examples include police officers and immigration officials demanding free sex from commercial sex workers, government officials taking bribes from people traffickers and so on (for many examples, see Darwin *et al.* 2003). We also have to deal with the reality that in many countries, new regulations or laws are simply an invitation to new forms of corruption.

What are the key shortcomings in government policy on health of international labour migrants?

Sending countries

Governments of 'sending countries' in the region, including the Philippines, Indonesia, Myanmar and Bangladesh, have seen overseas labour migration – both documented and undocumented – as a way of increasing income, and alleviating unemployment, without giving a great deal of attention to the well-being of migrant workers. Indeed, in some cases, they appear willing to 'trade off' the welfare of migrant workers against the greater foreign exchange to be earned by increasing the numbers of such workers. Embassies of the sending countries in the host countries frequently appear uninterested in doing much to assist migrant workers. In their defence, they often lack the staff and budget needed to do more; but their attitudes also appear to be conditioned by the low social status of the migrant workers concerned, and a recognition that the government they represent does not give high priority to the welfare of migrant workers. As Piper has put it, 'based on their class, gender and ethnicity, migrants are typically trapped between diplomatic concerns of elite and inter-governmental politics at both ends of the migra-tion chain' (unpublished communication).

In Hong Kong, for example, Filipino domestic workers, not their embassy, have organized themselves to press for better conditions. In con-trast, Indonesian maids have failed to organize, nor has their embassy done much, and as a result they have poorer conditions of work than Filipino domestic workers, who have gained one day's leave a week. Large numbers of Indonesian maids have fallen from high buildings in Singapore over the past few years, but for some time these cases did not appear to generate the degree of concern that might have been expected.

The Indonesian maid trade to Saudi Arabia has a long history, as does the abuse of many of these maids by their employers (Tirtosudarmo and Haning 1998). The Indonesian Embassy has long been aware of the abuse, as it is to the Embassy that many of those who manage to escape from abusive conditions flee for refuge. However, although it is possible that behind-the-scenes pressures have been applied, there is scant evidence of official concern by the Indonesian government about the ongoing abuses of the maids, who are bringing useful foreign revenue to Indonesia.

Abuse and cheating of Bangladeshi labour migrants returning to Dhaka and of Indonesian migrant workers at Cengkareng airport when they return from overseas have, at times, reached such a level of organization that failure by the authorities to act decisively to stamp them out can only be put down to a basic lack of concern about the welfare of labour migrants and/or collusion of officials in the abuses.

The need for contingency funds at embassies to cover the costs of hospital care for migrant workers involved in accidents is highlighted in a paper on Indian labourers in Lebanon (Gaur and Saxena 2003). The paper also highlights the problems of illegal workers, who were afraid to seek medical care because they feared deportation and arrest, highlighting the need for governments of both sending and receiving countries to cooperate in regulating the flow of migrants through legal channels.

Receiving countries

Governments of 'receiving countries', particularly in the case of illegal or undocumented migrants, are reluctant to give the same level of access to health services as they give to the local population. As Maruja Asis points out (Chapter 7, this volume), the status of unauthorized Filipino migrants to Sabah affects the conditions of their exit from their home country, the journey, their working and living conditions, and their return to their area of origin. In some ways, the reluctance to give the same level of services to migrants as to the often poor local population is realistic, given the resentments that are likely to arise if the local population think that migrants are getting resources denied to them. This was clearly evident in the reactions to the treatment of East Timorese refugees in West Timor after Timor Leste gained independence. But neglect of migrant health needs both disadvantages the migrant population and puts the host population at risk.

The problem is sometimes exacerbated by the different beliefs about disease and medical treatment between migrant and local populations, exacerbated by language problems and an inability to pay for medical attention

(Isarabhakdi 2004). The problem is sometimes further exacerbated by the near invisibility of certain groups of undocumented migrants; for example, Burmese domestic workers in northern Thailand, who live-in and are thus out of reach of most organizations (Toyota 2003).

More realistic approaches are needed to provide a basic level of health services to such undocumented migrants, but an even more basic need is for governments to facilitate an increase in the share of the movement which takes place through regular, documented channels. This could be done in the case of both Indonesian and Filipino labour migration to Sabah if the Indonesian and Philippines governments took more imaginative approaches to, for example, facilitating applications for work permits, and providing better transport services (Raharto 1999).

Sending–receiving country interactions

Illegal migration is a sensitive area in intergovernmental relations, as witnessed by the furore in both Indonesia and Malaysia at times when Malaysia threatens to repatriate large numbers of illegal workers. A crisis point was reached in early 2002, when widespread crime and rioting blamed on Indonesian workers led to a massive deportation of illegal migrants. The situation became chaotic in Nunukan, across the border from Sabah, where deported illegal Indonesian workers banked up, and disease and death occurred as a result. The issues caused a serious rift in Indonesian–Malaysian relations, but attention to the diplomatic rift appeared to overshadow the plight of the workers themselves (Darwin *et al.* 2003: 272–7).

How can government policy relevant to migration and health be improved?

The challenges to government policy are on two different levels. One can be met by more openness to the realities of the situation in which migrants find themselves, including health-seeking beliefs and practices among particular migrant groups, leading to more realistic and more effectively targeted policies in addressing these realities. Some governments in the region are unwilling to realistically face the reality of premarital sex, involvement of husbands in frequenting sex workers, and other unpalatable truths.[3] Even in Thailand, which has shown the way in open attitudes towards these issues and in addressing them, there are still myths surrounding the notion of HIV/AIDS being imported by foreign workers in border areas, and a reluctance to take the necessary steps to give this group effective access to services, despite the risks that such neglect poses to the local population.

But at another level, there are *structural factors* influencing migration and the health of migrants, including undocumented migration and within-country floating populations, and influencing the situation of migrant workers, which cannot be fully dealt with without attention to contextual problems (including, for example, wide differences in income levels, and the powerless position of migrant women factory workers and domestic servants). These contextual problems are much harder to deal with.

Xiang's (Chapter 9, this volume) discussion of China is relevant here. He shows that a series of institutional barriers serve to exclude large numbers of rural–urban migrants from the health care system – not intentionally, but as a logical outcome of the rural–urban dualist structure in China, and the system that is in place to provide health services. Many migrants enter the job market informally while the medical care system needs a formal employment relationship as a basis for operation. These institutional rigidities will not be easily overcome.

The motivation for dealing with these contextual problems has to be much broader than dealing with the health problems attendant on migration. It is clearly unrealistic to say, 'the whole thrust of your macro-economic policy is disadvantaging certain migrant populations and causing them health problems; therefore you need to rethink your macro-economic policies'. It is essential that a broader justification for modifying policy be presented to government; otherwise government will say, in effect, 'the ill-effects on migrant health that you are worried about may be an unfortunate by-product of our development strategy, but we will stick to our policies because they are going to bring greater prosperity to the bulk of our population'. There is a need to look for coalitions of interest, and policy synergies, where it can be shown that the benefits to migrant health will be linked to other benefits which might have stronger political appeal.

Perhaps one way to get governments to pay greater attention to the issues is to emphasize the 'multiplier effects' of ignoring them – for example, the spread of HIV/AIDS to the wider community, or the spread of tuberculosis or other untreated diseases through domestic servants' close contact with employers' families. Another way is to organize civil society to take the front running. This can both help deal with the health issue, and mobilize political support. In Japan, civil society has taken up the issue of the rights of migrant workers to access health services. In Thailand, NGOs such as Doctors of the World have developed a Partnership for Migrant Health with Mae Tao Clinic and the Thai Ministry of Public Health to increase access to health services for more than one million migrants living as unrecognized refugees in Thailand (Gaur and Saxena 2003: 15). In China, where the institutional barriers noted above seriously disadvantage migrant workers' search for health care, alternative approaches seem to have the potential to fill at least some of the needs. These alternative approaches include attention to migrant needs in government-funded community-based health systems, delivery of services and training by NGOs, and migrant-sponsored clinics which provide services more attuned to migrant life-styles and problems (see Xiang, Chapter 9, this volume). Such approaches can be introduced more speedily than those requiring formal policy change, mobilize social capital and build support for institutional changes.

Appropriate strategies for intervention

There are different possible ways governments could intervene in relation to the health problems of vulnerable groups within the population, including some categories of migrants and those affected by migration. One would be to take a generic approach – to determine that certain *categories* of people have greater health risks. It may not be necessary to specify migrants as a category in all circumstances – sometimes particular age groups or occupation groups will be a more effective focus. There is no need to flag all health problems associated with migration as migration problems per se if being a migrant is not really the core of the problem. In some situations, for example, the health problems of migrant commercial sex workers (CSWs) may result from their being CSWs, rather than from their status as migrants. Similarly, in the Philippines, maternal anaemia has been shown to be serious for both migrant and non-migrant mothers (Feranil, Chapter 6, this volume). In such a case, the focus needs to be on anaemic mothers, not on their migration status.

But where it *does* seem appropriate to target migrants as such, it may be necessary to focus on certain categories, for example, men moving without their families (India), concentrations of mining workers in remote areas, concentrations of commercial sex workers, or labour migrants about to leave for overseas. Another approach (which can be combined with the first approach) would be geographic – for example, looking for concentrations of high-risk migrant groups, or concentrated source areas of overseas migrant workers. It is well known that such concentrations do occur. Tangail district in Bangladesh, for example, is the source of large numbers of labour migrants to Singapore. Several hundred males have migrated to Singapore from one village there – Gurail – and almost every family in the village has a member who has spent time in Singapore (Rahman 2003: 89–90).

Effective intervention usually requires data. Governments need to give attention to the methodology of rapid appraisal reports on migrant health problems as they emerge, drawing on experience with other sectors such as agricultural forecasting. Sentinel surveillance surveys of HIV/AIDS need to pay more attention to migrant communities at risk. In addition, more research is needed to identify groups with high health risks associated with migration, including the migrants themselves, people in contact with them and left-behind communities. The health of 'left-behind' communities can benefit from remittances, but can also suffer through lack of physical support from absent family members. A key high-risk group who are given inadequate attention are the 'floating population' in China ('freelance immigrants' in Vietnam), who also have their parallels in Indonesia, Thailand and elsewhere. However, if these are left out of account in public health programmes, the HIV/AIDS control programmes and so on, this is going to place the whole programme at risk.

In the end, government policy cannot expect to cope effectively with all problems related to migration – a key issue, then, is how to *prioritize* what is important for policy to tackle. Civil society needs to be much more involved in seeking solutions. We can see this happening in many countries across Asia.

HIV/AIDS as a key area requiring government intervention

It has been claimed by UNAIDS and other groups working in the area that HIV/AIDS could be a problem of titanic proportions for China, as well as for countries such as Indonesia, where its spread seems to have been relatively restricted so far (UNAIDS/WHO 2003). It is therefore clearly an issue for governments to take very seriously, particularly given the likelihood that proven cases of HIV/AIDS tend to be the tip of the iceberg. At the same time, we must be realistic in admitting that, even bearing in mind the data uncertainties, HIV/AIDS has struck more severely in Thailand than in Indonesia or the Philippines, and that we do not fully understand why this has been the case. This uncertainty makes it difficult to determine priorities between certain more costly measures against HIV/AIDS and health measures needed in other priority areas.

Despite this uncertainty, there is no doubt that HIV/AIDS must be taken very seriously. Moreover, not all measures to combat it are expensive. Although HIV/AIDS has many links with migration, not all measures to counter it need to be directed specifically towards migrants. Government policy should include educational programmes on HIV/AIDS risks, condom use and so on among the population as a whole, but especially among high-risk migrant groups – overseas contract workers, seafarers, transport workers, the army, CSWs and so on – and among immigrant groups from neighbouring countries. There must also be enforcement programmes with brothels. There is a need for tough approaches to trafficking and child prostitution, as well as the enforcement of condom use in brothels. Other programmes should include linking NGOs dealing with overseas contract workers with those dealing with HIV, and peer education campaigns at places of entertainment.

Official action to counter HIV/AIDS needs to pay particular attention to border crossing points, because these tend to be foci for the spread of disease, including HIV/AIDS. Special approaches are needed due to the reality of illegal movement across such borders. Illegal migrants do not want to be interviewed, and also face language problems. Public health facilities in near-border areas must make special efforts to render themselves accessible to such migrants. Efforts are also needed to speed up the processing of applications to cross these borders more quickly, to minimize a bank-up of idle people at crossing points.

Policy implications related to labour migration

Given that migrant workers' contributions to the economies of both sending and receiving countries are frequently made at cost to the health and well-being both of the migrants and their families, migration policies that protect the basic human rights of migrant workers need to be ratified by sending and receiving nations. Policy guidelines developed by the International Labour Organization in the 1990 International Convention on Migrants' Rights, which include health rights, should be strictly implemented. In many cases where substantial labour migration flows exist between two countries, bilateral agreements for protecting the health of migrant workers are needed.

NGOs and migrant support groups of various kinds can play a role in pressing for the recognition of the health rights of labour migrants. NGOs have a particular role to play in assisting illegal migrants who face many obstacles in accessing health services. However, given the vulnerability of illegal migrants to many sources of exploitation which impact adversely on their health, the even more basic need is for international cooperation in curbing illegal migration. Nevertheless, even well-coordinated efforts in this direction can succeed only partially if the underlying determinants of illegal migration are not addressed. As long as vast inequalities exist between nations, poverty and unemployment will drive people to migrate, legally if this route is possible, but illegally if it is not.

Policy implications relating to left-behind women and families

As two chapters in this volume (Kuhn, Chapter 10; Roy and Nangia, Chapter 11) amply demonstrate, the relationship between out-migration of males and the health of left-behind women and families is complex. The increased earnings of successful migrants can ease the burden of paying for medical treatment for family members in the place of origin, but, because of economic distress, and risky practices engaged in by migrants, migration may endanger the health of spouses in particular. Better health education, especially about sexually transmitted diseases, is badly needed. Because government medical services may not be easily accessible to everybody in the villages, special efforts are needed to train traditional healers, traditional midwives, schoolteachers, and others respected in the community in sanitation and health issues related to reproductive tract infections/sexually transmitted infections (RTI/STI) and HIV/AIDS. Vehicles for dissemination of such knowledge would include radio, posters displayed in public places and on public vehicles, and local forms of entertainment. At the same time, because of the relative powerlessness of wives in such situations, it is crucial to influence the knowledge and behaviour of migrant males in order to lower the health risks to their spouses of their risky behaviour.

Health of ageing migrant populations

In countries with large permanent migrant populations, which are ageing over time (for example, in Australia and Singapore), there is a need to sort out the particular needs of these migrants for health care as they age. This will involve studying their family situation (for example, the availability of potential carers may be different between them and the local-born elderly), language problems and so on. Mental health problems may be of special concern among ageing migrants who have faced language and employment problems, separation from family, negative community attitudes and traumatic experiences prior to migrating. Programmes need to be designed with these special needs in mind.

Psychological stress and mental health issues among migrants

The prevalence and severity of mental health issues among migrants are likely to be affected by the pre-departure circumstances of the migrants, the particular circumstances of their move, and the conditions in their place of settlement. The problems are likely to be particularly dire in the case of many irregular or unauthorized migrants – pre-departure conditions such as poverty and armed conflict, lack of adequate health care for months or years during the travel itself, and/or trafficking or smuggling. Not surprisingly, severe mental health problems have been documented among some women trafficked into sex work (Darwin *et al.* 2003: 265–7).

Overall, however, Ho's study (2004) showed that in New Zealand at least, mental health levels among Asian migrants did not differ much from those of the general population. Many of the risk factors – language problems, failure to find employment, separation from family and community, negative public attitudes, and traumatic experiences prior to migration – are amenable to change, thus providing scope to lower mental health problems by effective interventions. The mental health system in countries such as Canada, Australia and New Zealand needs to be more responsive to the needs of Asian migrant service users.

Conclusions

We are dealing here with very complex matters, often lacking priority in a government's hierarchy of concerns. Effective policy requires a better understanding of the issues and the will to do something about them. Different levels of government have a role to play, from the international community through national governments to regional and local government. The migrants and their left-behind households are typically of low status, lacking a voice in elite politics, and also relatively invisible. In this situation of powerlessness they need advocates, and this is why NGO activism appears to be particularly important for migrant health and well-being. The more publicity that can be given to the problem, the more likely it is that effective responses will be seriously sought.

In order to underpin better policy on migrant health, more research is needed on many fronts. In general, the distinction between academic research and policy-related research is somewhat artificial in this, as in other fields. What is needed first and foremost is quality research. Any good research on migrant health issues is likely to have policy implications. This should not be taken as an excuse to conduct self-indulgent research on esoteric issues. The matters discussed in this book are too important for that, and the lack of knowledge on certain issues too glaring. A certain strategizing of research agendas is needed in order to ensure that important issues are addressed, and this can perhaps be best accomplished by mechanisms such as national workshops where researchers, policy-makers and planners come together to share their insights, concerns and priorities.

Notes

1 A good case in point is the example of the BIMP–EAGA agreement to help regularize travel from the Philippines to Sabah by opening a ferry service from Zamboanga to Sandakan. This did not greatly reduce the flow of undocumented workers, partly because potential migrants from Tawi Tawi to Sandakan (a boat trip of ten hours) still had to travel to Zamboanga, many hours in the opposite direction, in order to catch the ferry to Sandakan. The reason the Zamboanga–Sandakan ferry was not scheduled to call in at Tawi Tawi seems to be that the Philippines has a migration office in Zamboanga but not in Tawi Tawi.
2 Even so, illegal labour migration has the potential to sour relations between sending and receiving countries, as, for example, happened between Indonesia and Malaysia when Malaysia threatened (and began to take steps) to repatriate large numbers of Indonesian workers during the 1998 economic crisis.
3 In a study of young female migrant workers in garment manufacturing in Cambodia, more than half of them admitted to having premarital sex (Nishigaya 2003: 5)

References

Constable, N. (1997) *Maid to Order in Hong Kong: stories of Filipina workers*, Ithaca, NY: Cornell University Press.

Cox, D. (1997) 'The vulnerability of Asian women migrant workers to a lack of protection and to violence', *Asian and Pacific Migration Journal*, 6, 6: 59–75.

Darwin, M., Wattie, A.M. and Yuarsi, S.E. (eds) (2003) *Living on the Edges: Cross-border Mobility and Sexual Exploitation in the Greater Southeast Asia Sub-region*, Yogyakarta: Center for Population and Policy Studies, Gadjah Mada University.

Gaur, S. and Saxena, P.C. (2003) 'Indian migrant labour in Lebanon: a struggle for survival', Paper presented at the Migration and Health in Asia Conference, Bintan, Indonesia, Asian Metacentre for Population and Sustainable Development Analysis.

Hadiz, V. (1997) *Workers and the State in New Order Indonesia*, London: Routledge.

Ho, E., (2004) 'Mental health of Asian immigrants in New Zealand: a review of key issues', *Asian and Pacific Migration Journal*, 13, 1: 39–60.

Hugo, G. (1996) 'Labour export from Indonesia: an overview', *ASEAN Economic Bulletin* (Special Issue on labour migration in Asia), 12, 2: 275–98.

Hugo, G. (2000) 'Migration and women's empowerment', in H. Presser and G. Sen (eds) *Women's Empowerment and Demographic Processes*, Oxford: Oxford University Press.

Isarabhakdi, P. (2004) 'Meeting at the crossroads: Burmese migrants and use of Thai health care services', *Asian and Pacific Migration Journal*, 13, 1: 107–26.

Kassim, A. (1996) 'Alien workers in Malaysia: critical issues, problems and constraints', in C.M. Firdausy (ed.) *Movement of People Within and From the East and Southeast Asian Countries: Trends, Causes and Consequences*, Jakarta: Toyota Foundation and Southeast Asian Studies Program, Indonesian Institute of Sciences.

Lim, L.L. and Oishi, N. (1996) 'International labour migration of Asian women: distinctive characteristics and policy concerns', *Asian and Pacific Migration Journal*, 5, 1: 85–116.

Manning, C. (1998) *Indonesian Labour in Transition: An East Asian Success Story?*, Cambridge: Cambridge University Press.

Nishigaya, K. (2003) 'Young women workers in post-UNTAC Cambodia: structural adjustment, decent work deficit and sexual health risk', Paper presented at Migration and Health in Asia Conference, Bintan, Indonesia, the Asian Metacentre for Population and Sustainable Development Analysis.

Raharto, A. (1999) *Migrasi dan Pembangunan di Kawasan Timur Indonesia: isu ketengakerjaan* (Migration and Development in Eastern Indonesia: Labour Force Issues), Jakarta: PPT/LIPI.

Rahman, Md.M. (2003), 'Bangladeshi workers in Singapore: a sociological study of temporary labour migration', unpublished Ph.D. thesis, Department of Sociology, National University of Singapore.

Shah, N.M. and Menon, I. (1994) 'Violence against women migrant workers: issues, data and partial solutions', *Asian and Pacific Migration Journal*, 6, 1: 5–30.

Tirtosudarmo, R. and Haning, R. (1998) 'A needs assessment concerning Indonesian women migrant workers to Saudi Arabia', Jakarta: Center for Population and Manpower Studies, Indonesian Institute of Sciences.

Toyota, M. (2003) 'Health concerns of "invisible" foreign domestic maids in Thailand', Paper presented at Migration and Health in Asia Conference, Bintan, Indonesia, the Asian Metacentre for Population and Sustainable Development Analysis.

UNAIDS/WHO (2003) *AIDS Epidemic Update Dec. 2003*, UNAIDS/03.39E, pp. 18–22.

Index

Printed and bound by CPI Group (UK) Ltd, Croydon, CR0 4YY

01/11/2024

01782631-0004